AF412313

MANUFACTURING TECHNOLOGY
Volume 1: Engineering Materials

ELLIS HORWOOD SERIES IN MECHANICAL ENGINEERING

Series Editor: J.M. ALEXANDER, Professor of Mechanical Engineering, University of Surrey, and Stocker Visiting Professor of Engineering and Technology, Ohio University

The series has two objectives: of satisfying the requirements of post-graduate and mid-career engineers, and of providing clear and modern texts for more basic undergraduate topics. It is also the intention to include English translations of outstanding texts from other languages, introducing works of international merit. Ideas for enlarging the series are always welcomed.

Alexander, J.M.	**Strength of Materials: Vol 1: Fundamentals; Vol. 2: Applications**
Alexander, J.M., Brewer, R.C. & Rowe, G.	**Manufacturing Technology Volume 1: Engineering Materials**
Alexander, J.M., Brewer, R.C. & Rowe, G.	**Manufacturing Technology Volume 2: Engineering Processes**
Atkins A.G. & Mai, Y.W.	**Elastic and Plastic Fracture**
Beards, C.	**Vibration Analysis and Control System Dynamics**
Beards, C.	**Structural Vibration Analysis**
Beards, C.	**Noise Control**
Besant, C.B. & C.W.K. Lui	**Computer-aided Design and Manufacture, 3rd Edition**
Borkowski, J. and Szymanski, A.	**Technology of Abrasives and Abrasive Tools**
Borkowski, J. and Szymanski, A.	**Uses of Abrasives and Abrasive Tools**
Brook, R. and Howard, I.C.	**Introductory Fracture Mechanics**
Butkovskiy, A.G. & Pustyl'nikov, A.M.	**Mobile Control of Distributed Parameter Systems**
Cameron, A.	**Basic Lubrication Theory, 3rd Edition**
Collar, A.R. & Simpson, A.	**Matrices and Engineering Dynamics**
Cookson, R.A. & El-Zafrany, A.	**Finite Element Techniques for Engineering Analysis**
Cookson, R.A. & El-Zafrany, A.	**Techniques of the Boundary Element Method**
Ding, Q.L.	**Surface Modelling for CAD/CAM**
Dowling, A.P. & Ffowcs-Williams, J. E.	**Sound and Sources of Sound**
Edmunds, H.G.	**Mechanical Foundations of Engineering Science**
Fenner, D.N.	**Engineering Stress Analysis**
Fenner, R.T.	**Engineering Elasticity**
Ford, Sir Hugh, FRS, & Alexander, J.M.	**Advanced Mechanics of Materials, 2nd Edition**
Gallagher, C.C. & Knight, W.A.	**Group Technology Production Methods in Manufacture**
Gohar, R.	**Elastohydrodynamics**
Gosman, B.E., Launder, A.D. & Reece, G.	**Computer-aided Engineering: Heat Transfer and Fluid Flow**
Haddad, S.D. & Watson, N.	**Principles and Performance in Diesel Engineering**
Haddad, S.D. & Watson, N.	**Design and Applications in Diesel Engineering**
Haddad, S.D.	**Advanced Operational Diesel Engineering**
Hunt, S.E.	**Nuclear Physics for Engineers and Scientists**
Irons, B.M. & Ahmad, S.	**Techniques of Finite Elements**
Irons, B.M. & Shrive, N.	**Finite Element Primer**
Johnson, W. & Mellor, P.B.	**Engineering Plasticity**
Juhasz, A.Z. and Opoczky, L.	**Mechanical Activation of Silicates by Grinding**
Kleiber, M.	**Incremental Finite Element Models in Non-linear Solid Mechanics**
Leech, D.J. & Turner, B.T.	**Engineering Design for Profit**
Lewins, J.D.	**Engineering Thermodynamics**
Malkin, S.	**Materials Grinding: Theory and Applications**
McArthur, H.	**Motor Vehicle Corrosion**
McCloy, D. & Martin, H.R.	**Control of Fluid Power: Analysis and Design 2nd (Revised) Edition**
Osyczka, A.	**Multicriterion Optimisation in Engineering**
Oxley, P.	**Mechanics of Machining**
Piszcek, K. and Niziol, J.	**Random Vibration of Mechanical Systems**
Polanski, S.	**Bulk Containers: Design and Engineering of Surfaces and Shapes**
Prentis, J.M.	**Dynamics of Mechanical Systems, 2nd Editon**
Renton, J.D.	**Applied Engineering Elasticity**
Richards, T.H.	**Energy Methods in Vibration Analysis**
Richards, T.H.	**Energy Methods in Stress Analysis: with Introduction to Finite Element Techniques**
Ross, C.T.F.	**Computational Methods in Structural and Continuum Mechanics**
Ross, C.T.F.	**Finite Element Programs for Axisymmetric Problems in Engineering**
Ross, C.T.F.	**Finite Element Methods in Structural Mechanics**
Ross, C.T.F.	**Applied Stress Analysis**
Ross, C.T.F.	**Advanced Applied Stress Analysis**
Roznowski, T.	**Moving Heat Sources in Thermoelasticity**
Sawczuk, A.	**Mechanics of Plastic Structures**
Sherwin, K.	**Engineering Design for Performance**
Szczepinski, W. & Szlagowski, J.	**Plastic Design of Complex Shape Structured Elements**
Thring, M.W.	**Robots and Telechirs**
Tseung, A.C.C.	**Lotus 123 in Science and Technology**
Walshaw, A.C.	**Mechanical Vibrations with Applications**
Williams, J.G.	**Fracture Mechanics of Polymers**
Williams, J.G.	**Stress Analysis of Polymers 2nd (Revised) Edition**

MANUFACTURING TECHNOLOGY
Volume 1: Engineering Materials

J. M. ALEXANDER
DSc(Eng), PhD(Lond), DIC, FCGI, FIMechE, FIProdE, FIM, FRSA, MASME, FEng
Stocker Visiting Professor of Engineering and Technology
Ohio University, Athens, Ohio, USA

R. C. BREWER
BSc(Eng), AMIMechE, AMIEE
Department of Mechanical Engineering
Imperial College of Science and Technology
University of London, England

and

G. W. ROWE
MA, PhD, DSc, CEng, FIMechE, CPhys, FInstP, FRSA
Professor of Mechanical Engineering
University of Birmingham, England

ELLIS HORWOOD LIMITED
Publishers · Chichester

Halsted Press: a division of
JOHN WILEY & SONS
New York · Chichester · Brisbane · Toronto

This reworked and updated version
first published in 1987 by

ELLIS HORWOOD LIMITED
Market Cross House, Cooper Street,
Chichester, West Sussex, PO19 1EB, England
The publisher's colophon is reproduced from James Gillison's drawing of the ancient Market Cross, Chichester.

Distributors:

Australia and New Zealand:
JACARANDA WILEY LIMITED
GPO Box 859, Brisbane, Queensland 4001, Australia

Canada:
JOHN WILEY & SONS CANADA LIMITED
22 Worcester Road, Rexdale, Ontario, Canada

Europe and Africa:
JOHN WILEY & SONS LIMITED
Baffins Lane, Chichester, West Sussex, England

North and South America and the rest of the world:
Halsted Press: a division of
JOHN WILEY & SONS
605 Third Avenue, New York, NY 10158, USA

© 1987 J.M. Alexander, R.C. Brewer and G.W. Rowe/Ellis Horwood Limited

First published in 1962 as *Manufacturing Properties of Materials* **by D. Van Nostrand Co. Ltd.**

British Library Cataloguing in Publication Data
Alexander, J.M. (John Malcolm)
Manufacturing technology. —
(Mechanical engineering)
Vol. 1: Engineering materials
1. Materials
I. Title   II. Brewer, R.C.   III. Rowe, G.W.
IV. Series
620.1'1   TA403

Library of Congress Card No. 86–27426

ISBN 0–85312–025–0 (Ellis Horwood Limited)
ISBN 0–470–20815–5 (Halsted Press)

Printed in Great Britain by Unwins Bros., Woking

COPYRIGHT NOTICE
All Rights Reserved. No part of this publication may be reproduced, stored in a retrieval system, or transmitted, in any form or by any means, electronic, mechanical, photo-copying, recording or otherwise, without the permission of Ellis Horwood Limited, Market Cross House, Cooper Street, Chichester, West Sussex, England.

CONTENTS

CHAPTER 3   POLYMERS, ELASTOMERS AND REFRACTORY MATERIALS

CHAPTER 4   THE INFLUENCE OF PROPERTIES ON
PRODUCTION PROCESSES

CHAPTER 5   THE EFFECT OF PRODUCTION PROCESSES
       ON THE FUNCTIONING OF MATERIALS

PREFACE

This is the first of two volumes which together constitute a completely up-dated version of an original text by the late Roy C. Brewer and myself, entitled *Manufacturing Properties of Materials*, and first published by Van Nostrand, London, in 1963. Mr. Ellis Horwood, who was then the London publisher of Van Nostrand books, was later given exclusive rights to re-publication of this title when he formed his own publishing house in Chichester.

Roy Brewer died in 1965 at a relatively young age and it was a tragic loss to all who knew him. Some time after the second printing in 1968 I began trying to find an expert on machining to take Mr. Brewer's place and assist in up-dating the book. It actually took me 14 years to light upon someone suitable, qualified and in a position to undertake the task. It was my good fortune to find Geoffrey W. Rowe, Professor of Mechanical Engineering in the University of Birmingham, who agreed to help me in November of 1982 and I owe him a debt of gratitude for that. Dr. Rowe is an all-round expert in tribology, machining, metal forming, and materials science, so I am delighted that he agreed.

Finally, I owe almost as great a debt of gratitude to the Dean of Engineering at Ohio University, Dr. T. Richard Robe, who has encouraged me to up-date this work, and to write others with members of the Ohio University Engineering faculty, and has provided me with the necessary word-processing facilities. He also provided secretarial help from Miss Martha Dalton, who has skilfully and professionally processed the manuscript of the text. Much of the financial support for this purpose has come from the Stocker Fund, a generous endowment left to the College of Engineering and Technology, Ohio University, Athens, by one of its most famous alumni, the late Mr. C. Paul Stocker.

It is not a difficult task for me to dedicate this work. I dedicate it wholeheartedly to two great engineers, of quite different backgrounds, who have made different but equally valuable and necessary contributions to engineering and technology. They are the late Roy C. Brewer, a former colleague at the Imperial College of Science and Technology, University of London, my co-author for the first edition of this work; and the late C. Paul Stocker, that far-seeing engineer, philanthropist and benefactor who, in the true spirit of American generosity, provided endowment for a future he knew he would never live to see.

J. M. Alexander
Stocker Visiting Professor
College of Engineering and Technology
Ohio University
ATHENS, Ohio 45701-2979, U.S.A.
30th April 1986

Chapter 1

THE MECHANICAL AND PHYSICAL PROPERTIES OF MATERIALS

1.1 Introduction

There are several properties that determine the suitability of a material to a manufacturing process. In the first place, the strength of the material is very important, because it affects the ease with which it can be deformed into its required shape and its final ability to resist loads during service. Factors which either increase or decrease the strength of the material may be equally important. It may either be desirable to reduce its strength sufficiently to allow it to be 'pushed' into shape easily with the available machines, or alternatively to increase the final strength of the manufactured component and render it more serviceable. *Strength* is an imprecise generic term which may here be understood to indicate the tendency of a material to resist deformation.

A similar argument may apply to a rather more elusive property of a material – namely, its *ductility*. Ductility is usually understood to mean the ability of a material to accept large amounts of deformation (mainly tensile) without fracture. In some manufacturing processes a large value of this parameter will clearly be beneficial. Many metal working processes, for example, are limited only by the available ductility of the material being worked, so that the amount of deformation which can be imposed on the material has to be restricted to avoid fracture. There are, however, some manufacturing processes for which the antithesis of ductility is required. A suitable generic term for this property might be *brittleness*, and it is well known, for example, that certain brittle materials are much easier to machine or to shear than ductile materials.

It is mainly the interplay of properties such as strength and ductility during fabrication which has influenced the technology of production. Much of the skill of the modern artisan has been derived from the hard-won artistry of the old craftsman which resulted from empirical trial and error methods, sometimes extending over centuries. For example, it is well known that if a metal is heated it will become softer and easier to deform. If the speed of deformation is too great, however, this benefit will be lost, and the material may become either too hard or so brittle that fast deformation will lead to fracture. Clearly the occurrence and magnitude of such effects as these depend in some way on the microstructure of the material.

From this brief discussion may be seen the intention behind this book and its companion volume[1], both by the same authors. The aim is to try to indicate those properties of materials which are important both during and after manufacture, to see _why_ they are important and _how_ they influence the manufacturing process. It is clearly necessary to have more precise terms than 'strength' and 'ductility', and in this first chapter some of the standard mechanical tests will be considered to see whether it is possible to define such concepts with more precision. Of course, to do this it is really necessary to have some knowledge of the metallurgy of metals and the structure of non-metals, and also of the mathematical theory of the plasticity or rheology of ideal substances.

Once the various properties of importance in manufacture have been defined and understood, it is then possible to consider how this knowledge may be used to control the process and the product, and how these properties are affected by different production processes. In this way it should be easier to decide what is the best method of manufacture to suit a given component and given material. Of course, one might consider changing the shape or material of a manufactured component to suit the available production technique. Such questions are outside the scope of this book, and belong to the realms of design or production engineering. In the final analysis any successful manufacturing process must be economically sound and high priority should always be given to economic considerations. These should influence thinking from the outset — from the time a component is specified to fulfil a certain function for a certain lifetime — until its final inspection, testing, and guarantee. The whole process entails both design and production of the component, and these two books, which are concerned with a relatively small part of the complete problem, are intended as guides to the choice of material and process.

Finally, there are several physical and chemical properties which influence the choice and treatment of materials in manufacture. An example of such a property is that of thermal conductivity which will affect the flow of heat within the body of the material whilst it is being deformed and therefore its rate of cooling and hardening. A well-known example of an important chemical property is that of corrosion resistance. Its importance in the final product is obvious, and it may well be important during the manufacturing process too, because it can influence the formation of surface films which affect lubrication, or thermal and electrical conduction. A brief discussion of the more important physical properties is therefore included in this chapter.

1.2 Standard Mechanical Tests

It is very important to know the strength of a material, both for its eventual use and to determine the forces required to shape it. Since it is impracticable to test every article after it has been designed and made, several simple general tests are used to measure the mechanical properties of the identical stock.

1.2.1 *Tensile Tests*

The simplest and most widely accepted test uses a cylindrical or flat bar with enlarged ends. This is subjected to a steadily increasing tensile force, along its axis, and the extension of a gauge length is accurately measured, according to the appropriate standard[2]. The results commonly show the maximum tensile stress, the yield stress (Y) or proof stress, the percentage elongation to fracture and the reduction of cross-sectional area at fracture. In addition, the Young's Modulus of Elasticity (E) may be measured. Details are given in many standard text books[3,4,5].

It should be noted that the stresses recorded are *nominal*, obtained by dividing the actual load by the <u>original</u> area of the cross-section. For accurate calculations it is preferable to use the *true stress* σ_1, but this requires continuous evaluation of the <u>actual</u> area, which decreases during the test. Similarly the *nominal strain* depends on the original gauge length. A better measure is the *incremental true strain* $d\epsilon_1 = dl/l$, which can then be integrated for simple uniaxial stress states to give the logarithmic or *true strain* ϵ_1.

1.2.2 *Compression Tests*

It is often important for metal forming calculations to know the yield stress at much higher strains than can be obtained in tension. Axial compression of a short cylinder may be used, with suitable correction for the frictional resistance on the flat ends[6], but a more accurate result is obtained by the transverse plane strain compression of a well lubricated strip[5], illustrated in Fig 1.2.

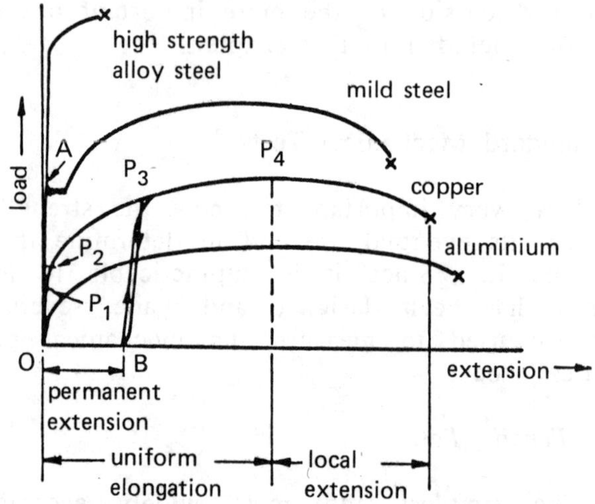

Fig. 1.1 Typical tensile test results
at slow speed and room temperature

1.2.3 *Hardness Testing*

Tensile and compressive tests are destructive of the sample, but it is often important to check the strength properties of stock material or finished components, without destroying them.

There are several types of hardness test for this purpose, which make only a small depression in the surface. The oldest and best known in U.K. are the test established by BRINELL[7], in which a standard ball, usually 10 mm diameter, is pressed into a metal under a load of 3000 kgf, and the Vickers test. The Brinell Hardness Number (HB10 or BHN) is then defined as the load in kgf divided by the curved surface area of the indentation in mm^2. It is found empirically, with some support from slip-line field theory, that the Brinell Hardness Number is directly related to the *uniaxial yield stress Y*. In fact, BHN $\simeq 2.9Y$, it being remembered that, to use this equation, Y must be in units of kgf/mm^2. If Y is in units of MN/m^2 (N/mm^2, or MPa), then BHN in its usual units of $kgf/mm^2 \simeq 3.3Y$. These tests are illustrated in Fig.

1.3.   The Vickers test uses a diamond pyramid with 136°
included angle (approximating to the contact angle of the
Brinell ball).

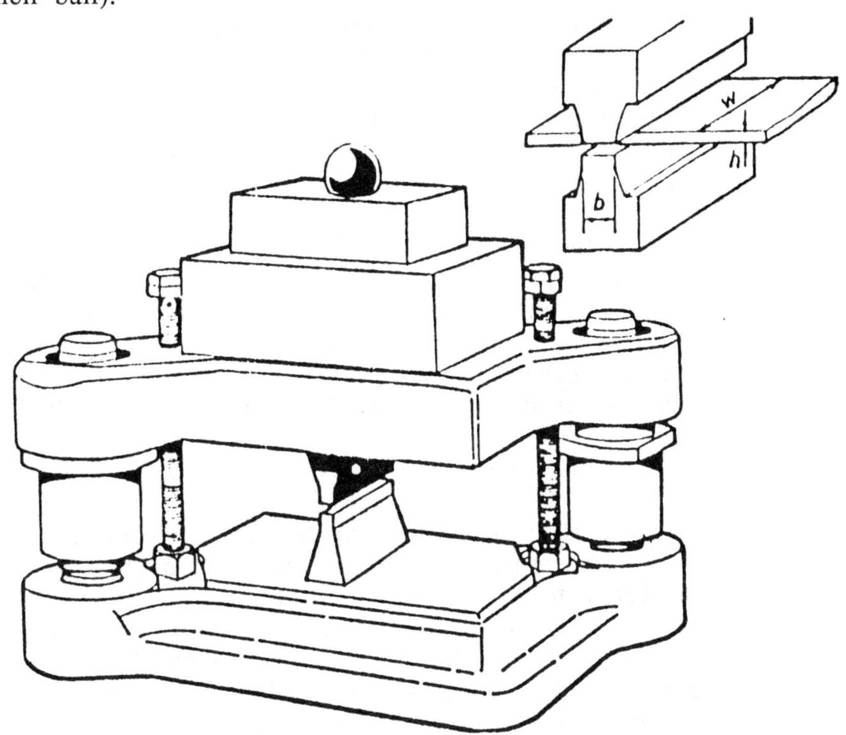

Fig. 1.2 Plane strain compression test

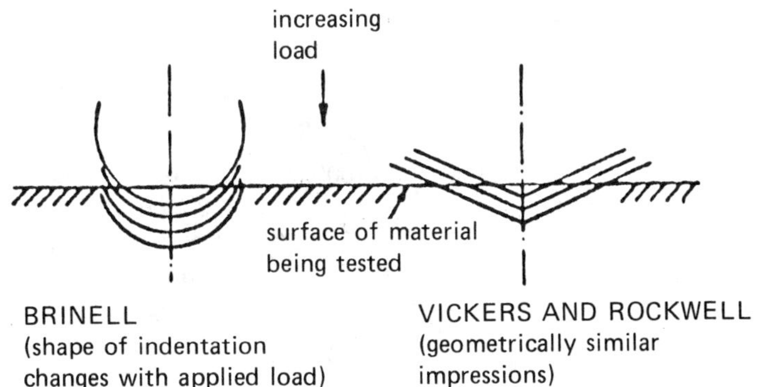

Fig. 1.3 Hardness tests

A much smaller load (2-120 kg) is applied and the area of the contact faces is determined from the diagonals of the very small indentation, as explained by TABOR[8] and discussed by ROWE et al.[9]. This test is frequently used for machined components while the Brinell test is mainly used on castings. For hardness values up to about 350 the HB and HV numbers are almost equal, but on harder materials the steel ball itself starts to deform and the reading is reduced. If a sapphire or tungsten carbide ball is used the equivalence extends further up the scale.

In the U.S.A., the Rockwell test[10] is favoured. In this test the depth of the indentation is measured whilst the load is still being applied, rather than the lateral dimensions. There is no direct correlation with Brinell and Vickers and the numbers have no specific dimensions. HR on the Rockwell C scale is approximately 0.1 HB over the range HB 200-400.

1.2.4 *Fatigue Tests*

It has been recognized for many years that static tensile or compressive testing is not adequate for components subjected to vibration. These can fail at very much lower stress levels, and there is a general relationship, due to Goodman, which shows the allowable oscillating stress level for a given mean stress, as shown in Fig. 1.4 for various surfaces produced by different modes of grinding.

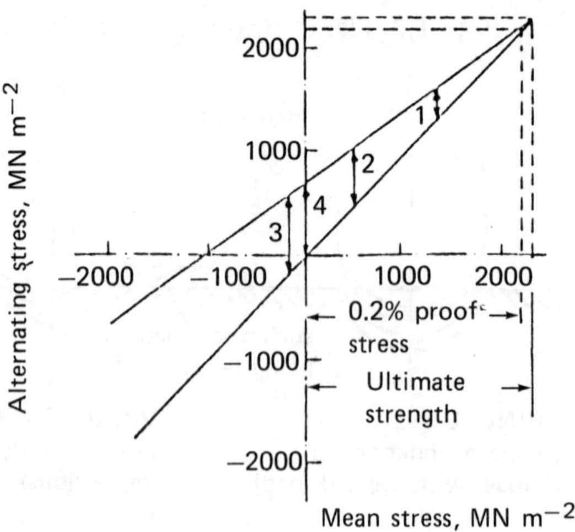

Fig. 1.4 A Goodman diagram for different ground surfaces
1 abusive; 2 conventional; 3 gentle; 4 annealed abusive

CADDELL[11] shows that fatigue testing needs considerable time, since each point on the final graph of applied stress S against the number N of cycles to failure requires a new specimen and N is usually between 10^6 and 10^8 (see Fig. 1.5).

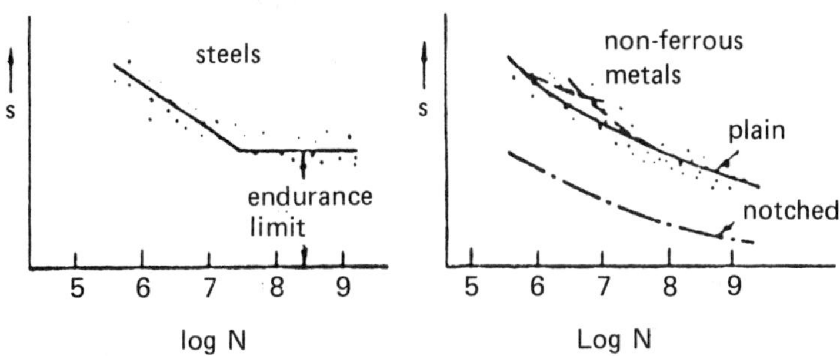

Fig. 1.5 Typical S-N curves

For many non-ferrous alloys the S-N curve falls steadily, but for steels there is often a levelling-off after some 10^6-10^7 cycles. If the stress does not exceed this endurance limit, the specimen will last indefinitely (see CRANE and CHARLES[12]). The cyclic loading is usually applied in rotating bending, or reverse bending, though for fatigue research a tensile or push-pull machine is preferred, with very carefully prepared surfaces.

VAN VLACK[13] shows that any surface cracks or other defects can seriously reduce the fatigue life, as can residual tensile stresses (see EL-HELIEBY and ROWE[14]).

High-stress low-cycle fatigue is sometimes important, especially in animal bone as pointed out by WAINWRIGHT et al.[15] and is potentially dangerous in aerospace components.

1.2.5 *Impact Testing*

Relatively brittle materials such as cast iron may fail under a single impact. Since it may be very important to avoid this type of fracture, impact tests have been devised in which a notched specimen is hit by a heavy pendulum. The energy absorbed is measured from the height of follow-through of the pendulum. The Izod test uses square section specimens (ROLLASON[3]) and the Charpy test, which is easier to conduct especially at elevated or lowered temperatures because the specimen is simply laid on the anvil, uses notched cylindrical bars. These tests are sketched in Fig. 1.6. The significance of temperature is that some alloys, especially common steels, become very susceptible to notch-brittleness at sub-zero temperatures, as pointed out about 30 years ago by TIPPER[16].

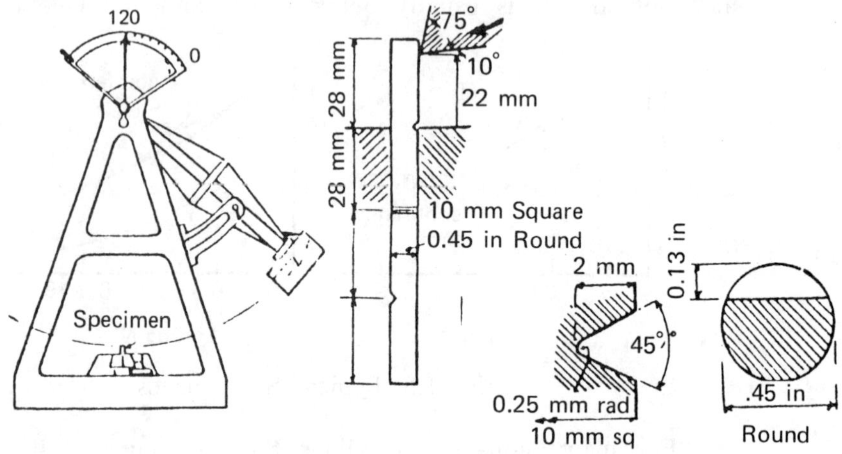

Fig. 1.6 Izod and Charpy specimens

1.2.6 *High-temperature Tests*

At high temperatures the plastic deformation of materials is dominated by diffusion processes which become evident above about 2/3 of the absolute melting temperature T_m, i.e. the homologous temperature T_H ($= T/T_m$) exceeds about 0.67 for hot working. The diffusion rate increases as the temperature increases, so the yield stress diminishes, but the latter also depends inversely on the strain rate as illustrated in Fig. 1.7. Tension, compression or hardness tests may be used at elevated temperature.

1.2.7 *Creep Tests*

An important feature of the hot tensile deformation of metals and alloys is that, at sufficiently high temperature, extension will continue at a very slow rate under very low loads.

This phenomenon of creep is very important in gas turbines and many other applications. Creep tests are conducted over long periods, typically 1000 hours (DIETER[17]).

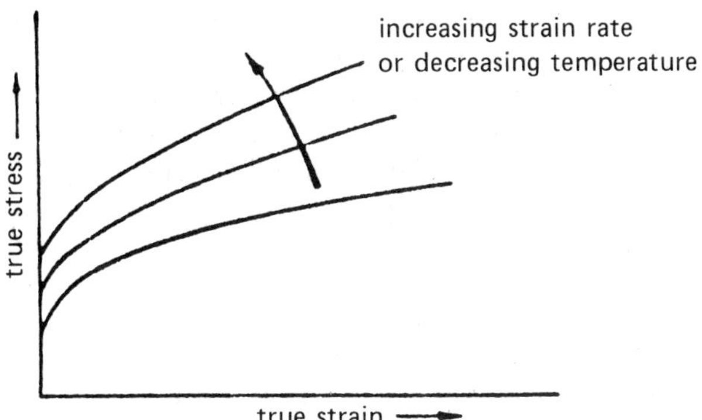

Fig. 1.7 The influence of strain rate
and temperature on yield stress

The results of a creep test can be presented graphically, generally exhibiting three distinct regimes as shown in Fig. 1.8. The secondary creep is usually recorded as the steady (minimum) creep rate. In the tertiary stage the decrease in cross-sectional area produces increasing stress and the creep accelerates to fracture.

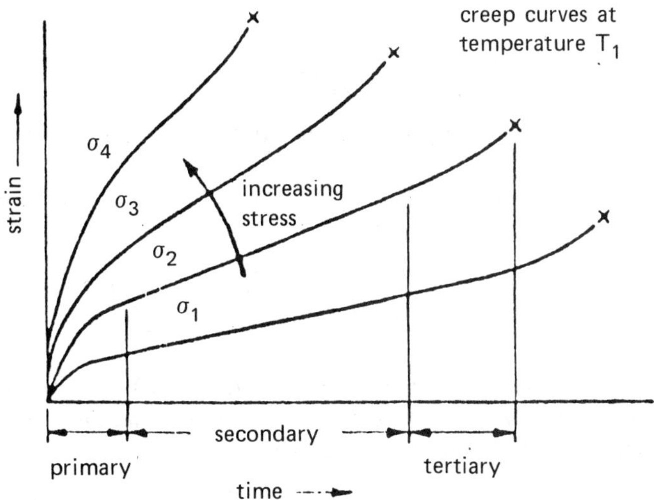

Fig. 1.8 Typical creep curves

Various creep equations have been suggested, notably that by ANDRADE[18], namely

$$\epsilon = A + B \ln t + C\, e^{\eta t} \tag{1.1}$$

Because of the length of time involved in creep testing, a shorter method is often used. Only approximate strain measurements are made during the test, the main purpose of which is to determine the time to rupture at a given stress and temperature. These stress-rupture tests can be further speeded up by testing a string of specimens in series in a long furnace.

1.2.8 *Fracture Toughness*

In recent years much attention has been given to the toughness of materials, which is related to the ease with which a crack, once started, will propagate. A simple view of this process, due to GRIFFITH[19] is that the opening of a crack releases elastic deformation energy but also requires the supply of surface energy to the two newly-created areas of crack surface. If, in a brittle material, the released strain energy U is sufficient for this, the crack will propagate.

For a plane crack of length $2c$ and an assumed cylindrical volume of $\pi c^2.1$ surrounding it, this condition is given by:–

$$U = \frac{\sigma^2}{2E}\pi c^2 \geqslant 2\gamma.2c \qquad (1.2)$$

where σ is the applied stress and γ is the surface energy per unit length of crack.

This leads to the definition of a critical stress at which the crack may propagate:-

$$\sigma_c \simeq \sqrt{\frac{4E\gamma}{\pi c}} \simeq \sqrt{\frac{E\gamma}{c}} \qquad (1.3)$$

Typical values for iron are $E = 210$ kN/mm^2, $\gamma = 2 \times 10^3$ N/mm and $c_0 = 2.5 \times 10^{-7}$ mm (the smallest possible crack, equated to one atomic spacing). Then:-

$$(\sigma_c)_{max} = 41 \times 10^3 \text{ kN/mm}^2 \text{ (GPa)} \simeq \frac{E}{5} \qquad (1.4)$$

Strengths approaching this value can be recorded for metallic "whiskers" which are free from mobile dislocations.

Metals in their common form are usually ductile, so further energy p will be absorbed in plastic deformation. Equation 1.3 should then be written to include this factor and any other energy loss, such as acoustic emission.

$$\sigma_c \simeq \sqrt{\frac{E(\gamma+p)}{c}} \qquad (1.5)$$

This makes the material less brittle, or tougher. The real situation is more complex, but a Strain Energy Release Rate G can be determined from an experimental Crack Opening Displacement (COD) test.† The force necessary to extend the crack from length c_1 to c_2 is determined (Fig. 1.9), from which the value of G, per unit area, can be found for a specimen of large width w, from the following equation:-

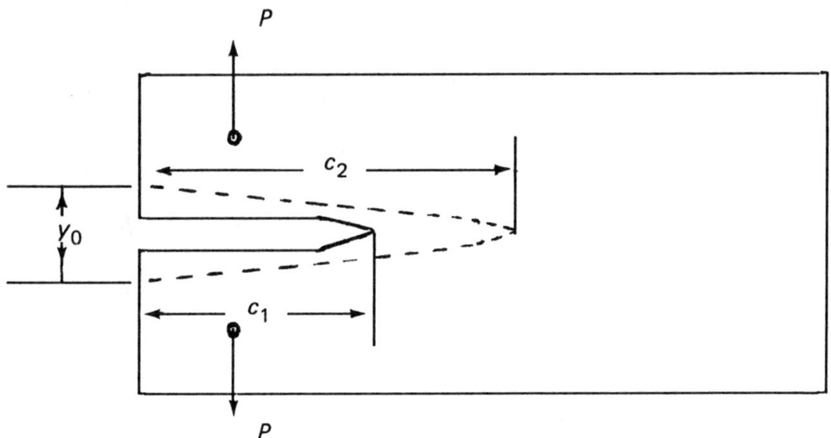

Fig. 1.9 COD determination

$$G = -\frac{dU}{dA} = -\frac{dU}{da}\cdot\frac{1}{w} = -\frac{y_0}{2w}\cdot\frac{\Delta P}{\Delta a} \qquad (1.6)$$

where y_0 is a pre-selected mouth opening of the crack. At a critical value G_c, the crack will propagate, as in Equation 1.5:-

$$\sigma_c = \sqrt{\left[\frac{EG_c}{c}\right]} \qquad (1.7)$$

The real situation is often complex and reference is made to the stress field, described by a Stress Intensity Factor K. This is not the stress concentration at a notch or other change in section, but describes the stress distribution and can be calculated for special conditions.

†{Sometimes referred to as the Crack <u>Tip</u> Opening Displacement (CTOD)}.

The critical value K_c for the growth of a crack in a thin sheet under *plane stress* is determined by the shape and size of the plastic zone around the crack and is known as the Fracture Toughness. This decreases as the thickness of a test plate is increased, having a minimum value K_{1c} for *plane strain conditions* because of the constraint imposed on the plastic deformation zone of a thick plate.

K_{1c} is therefore a material property from which the value of K_c for any other known configuration can be found. As described by HERTZBERG[20] the "toughness" K_{1c} is determined from the experimental measurement of force required to extend the crack by a certain length. For a given COD, the work required will depend upon the length of the original crack and its extended length.

Table 1.1 gives some comparative values:-

TABLE 1.1

Comparative fracture toughness values

	Strength MPa	Toughness MPa.m$^{\frac{1}{2}}$
Medium carbon steel	260	55
Pressure-vessel steel	500	200
Aluminium Alloy 2024-T4	350	55
7075-T6	460	40
7178-T6	560	23
Titanium alloy Ti-6Al-4V	830	55
Ti-4Al-4Mo-4S	1095	35
Polymethacrylate	30	1
Polycarbonate		2.2
Glass		0.4

Recently, ATKINS and MAI[21] have written a comprehensive text on both elastic and plastic fracture, giving considerable attention to the problems of fracture in materials subjected to large deformation plastic flow, which is therefore of great interest in the context of metal working processes.

1.2.9 *Plastic Anisotropy*

In sheet metal forming it is important to recognize that the properties of rolled sheet may differ substantially in the rolling and transverse directions, as well as in the thickness direction.

This feature can be measured in terms of the so-called *r*-value, which is the ratio of the transverse to the longitudinal strain in a tensile test on wide, flat strip using techniques described by HOSFORD and CADDELL[22]. Volume is always conserved in plastic deformation, so the thickness strain is also dependent upon *r*.

1.3 The Basic Stress-Strain Curve

1.3.1 *General Concept*

Consideration of the standard mechanical tests described above reveals that it should be possible to predict the behaviour of any given metal from the results of a single test, such as the tensile test, for example. Unfortunately, it is not possible to predict the precise stress-strain pattern which will exist in any given situation.

Nevertheless, the general concept that any given material possesses a basic stress-strain curve is useful, expecially in predicting the behaviour of the material when subjected to plastic flow. In most metal-working processes the stresses are predominantly compressive, and fracture does not occur very readily. Thus, although a tensile specimen will fracture after only about 40 per cent extension in the tension test, it is possible to roll or forge the same specimen until it has extended many times its original length, simply because the stresses in rolling or forging are predominantly compressive. During the deformation the metal must have been strain hardened, and the question arises as to how to determine the stress-strain curve at large strains. The results of the tensile-test are useful only up to relatively small strains, even if measurements are made right up to fracture. Also the method of defining strain for such large deformations is important; this is discussed in more detail in the companion volume[1].

There is ample evidence to show that the stress-strain curve for a given metal in uniaxial tension corresponds closely with the curve for a metal in uniaxial compression, provided the true stress (load/current cross-sectional area) and true strain (logarithmic strain) are used as co-ordinates (see Section 1.2.1). Clearly then, to obtain a stress-strain curve up to the large strains required, the best method is to test a specimen of the material in compression, using either the axial compression of a short cylinder or the plane strain compression test illustrated in Fig. 1.2. In the plane strain compression test the strain is substantially zero in the width direction of the specimen, due to the constraint of the material outside the platens (see Fig. 1.2). Since in simple uniaxial tension or compression there is no such restraint on the sideways movement of the material, more general definitions of stress and strain are required. These are the *effective* or *generalized* stress or strain.

As shown elsewhere[1], the simplest definitions of these quantities in terms of the principal stresses or strain increments are, using the yield criterion of Maxwell (or von Mises), as follows:-

Effective stress

$$\bar{\sigma} = \frac{1}{\sqrt{2}} \sqrt{(\sigma_1-\sigma_2)^2+(\sigma_2-\sigma_3)^2+(\sigma_3-\sigma_1)^2} \qquad (1.8)$$

Effective Strain Increment

$$\delta\bar{\epsilon} = \sqrt{\frac{2}{3}\left[\delta\epsilon_1{}^2+\delta\epsilon_2{}^2+\delta\epsilon_3{}^2\right]} \qquad (1.9)$$

In the plane strain compression test, $\delta\epsilon_2 = 0$ and therefore $\sigma_2 = \frac{1}{2}(\sigma_1+\sigma_3)$, as shown elsewhere[1]. Also, since the volume remains substantially constant during plastic flow, $\delta\epsilon_1+\delta\epsilon_2+\delta\epsilon_3 = 0$ and $\delta\epsilon_3 = -\delta\epsilon_1$.

Substituting these values in Equations 1.8 and 1.9 gives the following relationships between effective stress and strain, and <u>actual</u> stress σ_1 and strain ϵ_1 in the plane strain compression test:-

$$\bar{\sigma} = \frac{\sqrt{3}}{2}\sigma_1 \qquad (1.10)$$

$$\bar{\epsilon} = \frac{2}{\sqrt{3}}\epsilon_1 \qquad (1.11)$$

In uniaxial tension or compression, $\bar{\sigma} = \sigma_1$ and $\bar{\epsilon} = \epsilon_1$, σ_1 and ϵ_1 being the <u>actual</u> stress and strain in the specimen.

Thus, to derive the stress-strain curve in uniaxial tension or compression from the stress-strain curve obtained from a plane strain compression test (σ_1 vs. ϵ_1), it is necessary to divide the ordinates of the curve by the factor $2/\sqrt{3} \triangleq 1.155$ and also to multiply the abscissae by the same factor.

In the case of high temperature testing, where the rate of straining is of paramount importance, the definition of effective strain rate is obtained simply from the expression for the incremental strains (Equation 1.9) as:-

$$\bar{\varepsilon} = \sqrt{\frac{2}{3}\left[\bar{\varepsilon}^2 + \bar{\varepsilon}^2 + \bar{\varepsilon}^2\right]} \qquad (1.12)$$

1.3.2 *Constitutive Equations*

A discussion about constitutive equations has been given by ALEXANDER[23]. Approximate ranges of homologous temperature T_H and strain rate were investigated for many metal forming processes and are summarized in Table 1.2.

TABLE 1.2
Typical ranges of T_H and strain rate

Process	Range of T_H	Approximate range of strain rate (s^{-1})
Casting	0·9-1·0	0·0-0·4
Hot forging	0·7	100-200
Hot extrusion	0·7	10-20
Cold forging and extrusion	0·16-0·32	10-100
Warm forging	0·55	10-100
Creep forming	0·63+	$10^{-2}-10^{-4}$
Sheet metal forming	0·16-0·32	10-100
Powder forming	Not known	Not known
Deposition forming	0·9	200-1000
Machining, shearing	0·16-0·32	$\geqslant 1000$
Explosive forming	0·16-0·32	100-1000
Hot rolling	0.7	40-100
Cold rolling	0.16-0.32	100-1000

It may be concluded from this recent paper that one possible constitutive relation which might be useful for representing the dependence of flow stress $\bar{\sigma}$ upon effective strain $\bar{\varepsilon}$, effective strain rate $\bar{\varepsilon}$ and absolute temperature T is:-

$$\bar{\sigma} = D\bar{\varepsilon}^{n_1}\bar{\varepsilon}^{n_2}\exp\left(Q/RT\right) \qquad (1.13)$$

where Q is an activation energy
     R the universal gas constant
     D, n_1 and n_2 constants

Since Equation 1.13 gives a zero value for $\bar{\sigma}$ if either $\bar{\epsilon}$ or $\bar{\dot{\epsilon}}$ is zero, which is certainly <u>not</u> the situation in practice, a much more realistic equation is:-

$$\bar{\sigma} = \sigma_0(1+B\bar{\epsilon})^{n_1} \times (1+D\bar{\dot{\epsilon}})^{n_2} \exp (Q/KT) \qquad (1.13a)$$

A different approach, more geared to the dislocation mechanics of physical metallurgy, is that of ASHBY and FROST[24] who have developed *deformation mechanism maps* and *fracture mechanism maps*, typically as depicted in Fig.1.10.

The strain rate of a plastically deforming crystalline solid is dependent upon a number of basic atomic processes such as: dislocation motion, diffusion, grain boundary sliding, twinning, or a phase transformation. Such processes can combine to give at least 12 distinctive deformation mechanisms, e.g. yielding (or slip), power law creep, Nabarro-Herring creep, Coble creep, super-plastic flow, etc. etc.

Deformation mechanism maps for any polycrystalline material can be constructed in stress-temperature space, showing the area of dominance of each flow mechanism for each of which a rate equation exists, linking shear strain rate $\dot{\gamma}$ to shear stress τ, temperature T and to 'structure' (including all parameters describing its atomic structure, e.g. bonding, crystal class, defect structure, grain size, dislocation density and arrangement, solute or precipitate concentration, etc.). An example of such a map is shown in Fig. 1.10 on which typical equations have been shown, relating the normalized shear stress τ/μ, where μ = shear modulus, to homologous temperature. KELLY[25], has shown that τ cannot exceed τ_{ideal} ($\simeq\mu/20$), shown by the line marked 'ideal strength' in the figure, which is for pure nickel, but polycrystalline metals deform at stresses much less than this by dislocation glide and/or movement of vacancies. Examples of other maps and methods of using them are described by ASHBY and FROST[24]. In some forming processes, where very large strain rate ranges may be encountered, the description of steady state behaviour given in Equation 1.13 is inadequate. At very high stress, an exponential stress dependence on strain rate is more appropriate and the power law and exponential variation can be combined as:-

$$\bar{\dot{\epsilon}} = A \ \{\sinh (a\bar{\sigma})\}^{m} \exp (-Q/RT) \qquad (1.14)$$

This relationship has been shown to give a good fit with

experimental data for aluminium alloys over many orders of magnitude in strain rate.

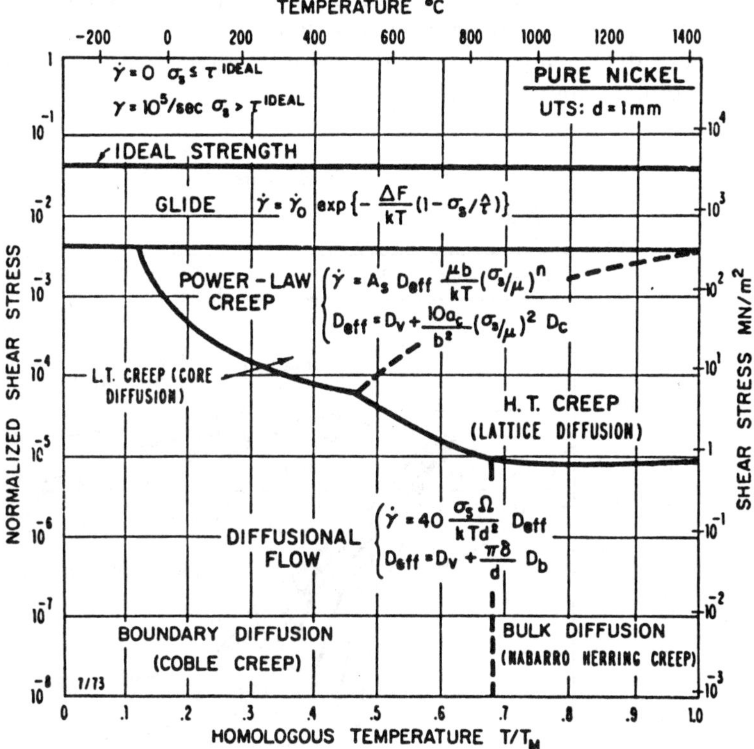

Fig. 1.10 The fields in which a particular mechanism of flow is dominant. Boundaries are found by equating flow rates

However, the description is still of 'steady state' behaviour and there appears to be no good evidence that it can be used as an *equation of state*† relationship when stress or temperature change sharply. This problem has been approached in two ways.

One procedure is to define an equation of state which contains an internal variable describing the structure of the material. If the variation of this internal variable with change in stress, temperature and time can be established, an adequate constitutive equation may result. Numerous attempts have been

†An *equation of state* implies a fixed one-to-one relationship between dependent and independent variables, e.g. as for the equation relating the pressure, volume, and absolute temperature of ideal gases in thermodynamics, viz. *PV = RT*.

made to provide such a description but they all suffer from the disadvantage that large numbers of sophisticated tests have to be conducted to establish the internal variables. Thus not only constant stress and constant strain rate tests are required but also complex relaxation and other stress and temperature cycling experiments.

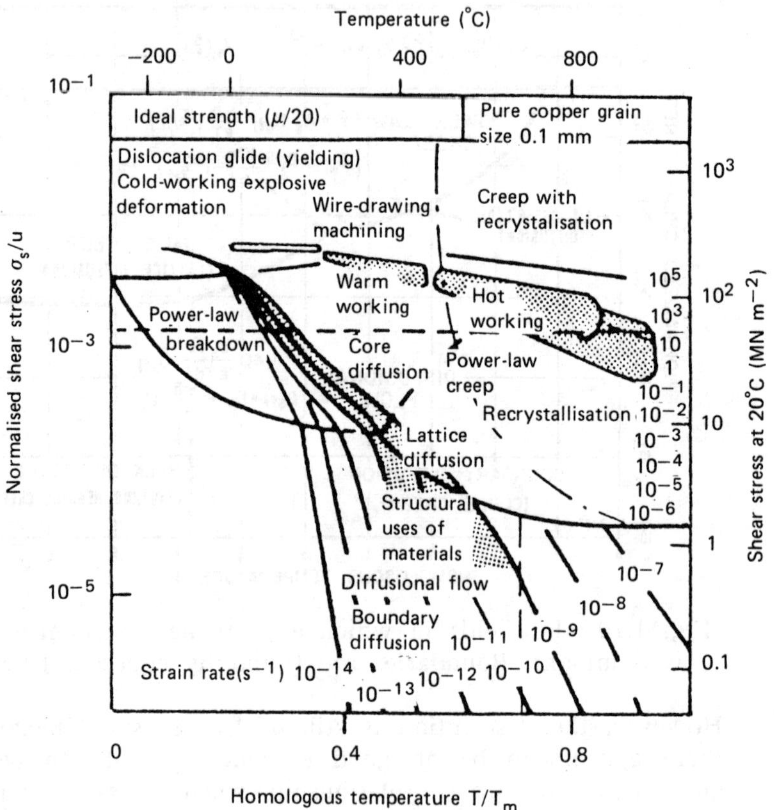

Fig. 1.11 Forming operations on a
deformation map for pure copper

Ashby's deformation and fracture mechanism maps have made an important contribution to the way of describing and illustrating both deformation and fracture phenomena. It is possible to delineate various regions for different deformation processes on a deformation mechanism map as illustrated in Fig. 1.11 (for pure copper). It is worth noting here that Ashby plots curves of normalized shear stress vs. homologous temperature, for various strain rates, derived from dislocation mechanisms. The same approach can be used for fracture.

1.3.3 *Modelling of Dynamic Material Behaviour in Hot Deformation*

Another (similar) approach has been adopted by H. GEGEL and his co-researchers, PRASAD et al.[26]. This approach, however, is based on thermodynamic considerations and does not rely on the microstructural behavioural description of the material. It is 'similar' only in so far as the final presentation is of areas in which deformation is possible without fracture or defects, and it is possible to identify optimum regions in which deformation is most efficient. This approach seems to be superior to that of Ashby in that it is more general (not necessarily tied to metals) and is easier to understand. Attention is drawn towards the instantaneous power being dissipated, viz.:-

$$\overline{\sigma}\overline{\epsilon} = \int_0^{\overline{\epsilon}} \overline{\sigma} \, d\overline{\epsilon} + \int_0^{\overline{\sigma}} \overline{\epsilon} \, d\overline{\sigma} \qquad (1.15)$$

or

$$P = G + J \qquad (1.16)$$

where the power dissipated by plastic work is denoted by the area G under the curve in Fig. 1.12 and designated the *dissipator content* whilst the area J above the curve is termed the *dissipator co-content* and is related to the structural (metallurgical or non-metallic) mechanisms which occur dynamically to dissipate power.

From Equation 1.15 it follows that, at any given temperature and effective strain, the partitioning of power between J and G is given by

$$\left[\frac{\partial J}{\partial G}\right]_{T,\overline{\epsilon}} = \left[\frac{\partial \ln \overline{\sigma}}{\partial \ln \overline{\epsilon}}\right]_{T,\overline{\epsilon}} \qquad (1.17)$$

which is simply the strain rate sensitivity of a material conventionally evaluated as:-

$$m = \frac{\partial(\log \overline{\sigma})}{\partial(\log \overline{\epsilon})_{T,\overline{\epsilon}}} \simeq \frac{\Delta(\log \overline{\sigma})}{\Delta(\log \overline{\epsilon})_{T,\overline{\epsilon}}} \qquad (1.18)$$

(Note: $m = n_2$ of the discussion relating to Equation 1.13.)

Determining the values of m as a function of effective strain rate for a material and assuming that the exponential equation:-

$$\overline{\sigma} = A\overline{\dot{\epsilon}}^{m} \qquad (1.19)$$

is valid for each value of m, the values of J are then calculated as a function of effective strain rate at a given T and $\overline{\dot{\epsilon}}$. A piece-wise quadratic fit is used to determine the various values of m over the range of strain rate used in the tests.

Thus, the power dissipated in changing the material structure is:-

$$J = \int_0^{\overline{\dot{\epsilon}}} \overline{\dot{\epsilon}}\,d\overline{\sigma} = \int_0^{\overline{\sigma}} (\overline{\sigma}/A)^{1/m}d\overline{\sigma} = \frac{1}{A^{1/m}} \cdot \frac{\overline{\sigma}^{(1/m+1)}}{(1/m+1)} = \frac{\overline{\dot{\epsilon}}.\overline{\sigma}.m}{m+1} \qquad (1.20)$$

Gegel and his colleagues show elsewhere that m must lie between 0 and 1 if the material deformation is not to become unstable. For values of $m < 1$ the curve of $\overline{\sigma}$ versus $\overline{\dot{\epsilon}}$ appears similar to that shown in Fig. 1.12(a). When $m = 1$, the curve is a straight line giving a maximum dissipator co-content (of area $J_{max} = \overline{\dot{\epsilon}}.\overline{\sigma}/2$), as shown in Fig. 1.12(b).

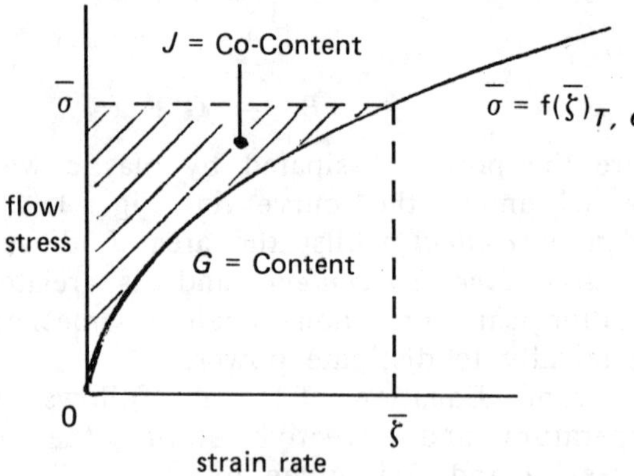

Fig. 1.12(a) Schematic representation of G content and J co-content for workpiece having a constitutive equation represented by curve $\overline{\sigma} = f(\overline{\dot{\epsilon}})$. Total power of dissipation is given by the rectangle.

Hence, it seems sensible to define an efficiency of deformation in terms of a non-dimensional parameter, in line with other stability criteria, namely

$$\eta = \frac{J}{J_{max}} = \frac{2m}{m+1} \qquad (1.21)$$

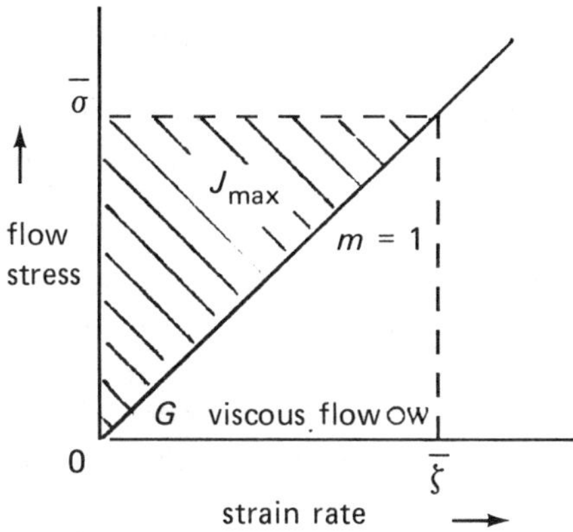

Fig. 1.12(b) Schematic representation showing J_{max}
which occurs when the strain-rate sensitivity
(m) of the material is equal to one

Ti-6242, BETA, 0.6 STRAIN

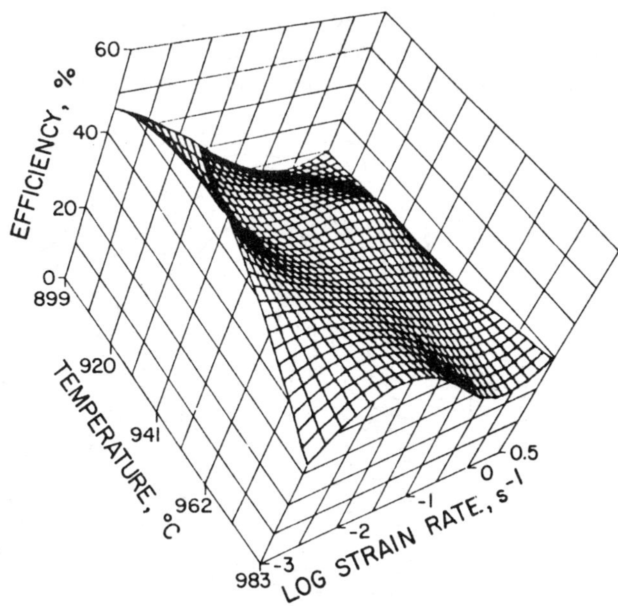

Fig. 1.13 Three-dimensional plot showing variation of
efficiency of dissipation with temperature and strain-rate for
Ti-6242 preform at 0.6 strain

These researchers then proceed to determine curves of $\bar{\sigma}$
versus $\bar{\varepsilon}$ for gas turbine disc materials such as Ti-6242 at
various temperatures and strains, by cross-plotting and

interpolation of data obtained from conventional stress-strain compression testing up to high strain values. A three-dimensional plot of efficiency η vs temperature T and logarithmic strain rate $\ln\bar{\dot\epsilon}$ can then be constructed and two-dimensional maps showing η contours derived for various values of effective strain $\bar\epsilon$ as shown typically in Figs. 1.13 and 1.14.

It is also possible to delineate regions on these maps where fracture or defects are most likely to occur as shown in Fig. 1.15 and this is research currently being undertaken by the group under Gegel's leadership. For example, the parameter $\sigma_m/\bar\sigma$ (where σ_m = the mean or hydrostatic stress) can be determined everywhere throughout the volume of material being deformed (e.g. by finite element methods) and it must be ensured that it does not exceed algebraically a certain limiting (compressive) value, typically -2/3 for extrusion at the exit to avoid internal cracking (centre burst). This parameter, taken in conjunction with dynamic recrystallization, phase transformations, spheroidization of acicular structures, precipitation mechanisms, edge cracking and local kinking of β-platelets in Ti-6242 can all contribute to prediction of zones on these maps which should be avoided.

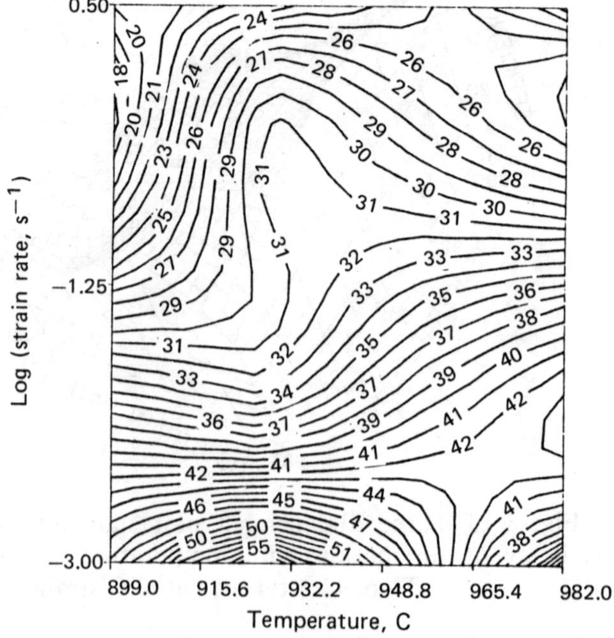

Fig. 1.14 Map showing constant efficiency contours in strain rate-temperature frame for Ti-6242 β-preform at 0.6 strain

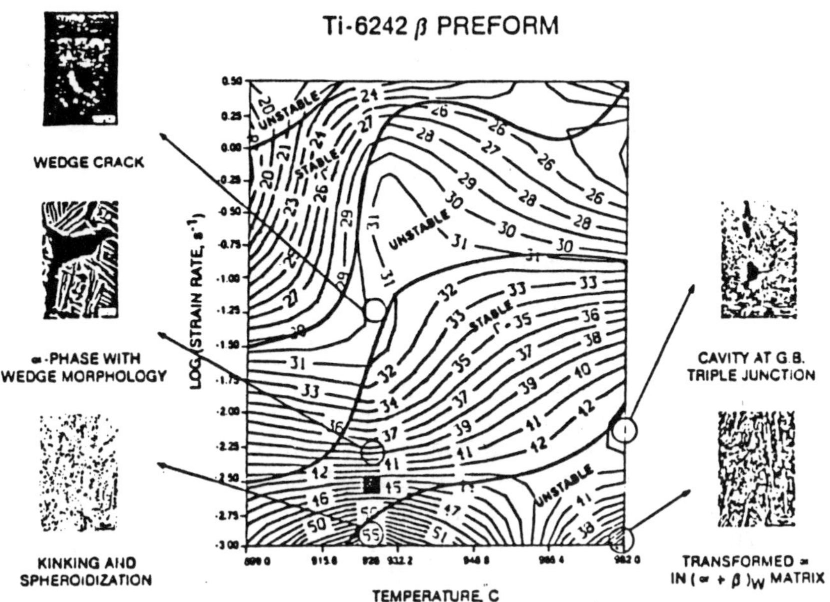

Fig. 1.15 Processing map for Ti-6242 β microstructure
with stable regions identified (courtesy of
Wright Patterson Air Force Base)

It is very interesting to observe that the predicted optimum
forging conditions of both $\alpha + \beta$ and β preform microstructures of Ti-6242 are 927°C and 10^{-3}s^{-1}, corresponding
almost exactly with the creep-forming conditions which have
been arrived at in practice by laborious and expensive trial
and error methods.   More recently, by applying Lyapunov
function stability criteria (well-known in control theory, e.g. as
described by SCHULZ and MELSA[27]) to material processing
variables such as m and introducing an analogous entropy
coefficient s related to stress σ and absolute temperature T by
the equation:-

$$s = \frac{1}{T} \frac{\partial(\log \bar{\sigma})}{\partial(1/T)}\bigg|_{\bar{\epsilon},\bar{\dot{\epsilon}}} \qquad (1.22)$$

Gegel and his colleagues have pointed out that:-

$$s = s_{\text{sys}}/s_{\text{app}} \qquad (1.23)$$

where s_{sys} = the rate of entropy production in the system
and s_{app} = the rate of entropy input into the system.
     The analogy between m and s can best be seen by
comparing Equation 1.18, viz.:-

$$m = \frac{\partial(\log\bar{\sigma})}{\partial(\log\bar{\epsilon})_{T,\bar{\epsilon}}} \simeq \frac{\Delta(\log\bar{\sigma})}{\Delta(\log\bar{\epsilon})_{T,\bar{\epsilon}}} \qquad (1.18)$$

with Equation 1.22.

Using the Lyapunov function stability criteria then leads to the condition that, in stable regions, the two second order partial differential coefficients of m and s with respect to log $\bar{\epsilon}$ should be <u>negative</u>, viz.:-

$$\frac{\partial m}{\partial(\log\bar{\epsilon})} < 0 \; ; \; \frac{\partial s}{\partial(\log\bar{\epsilon})} < 0 \qquad (1.24)$$

When the rate of change of the slope of a function is negative, this implies that the function itself is a maximum. For the material parameters m and s, therefore, the condition for stability is that they should both be as large as possible, within the obvious limitations that neither can exceed unity. (Because $m = 1$ corresponds with Newtonian fluid viscosity and $s = 1$ corresponds with the rate of entropy production in the system being equal to the rate of entropy being put into the system).

These two conditions are used in the development of processing maps of the type shown in Figs. 1.15 and 1.16, as reported by GOPINATH[28].

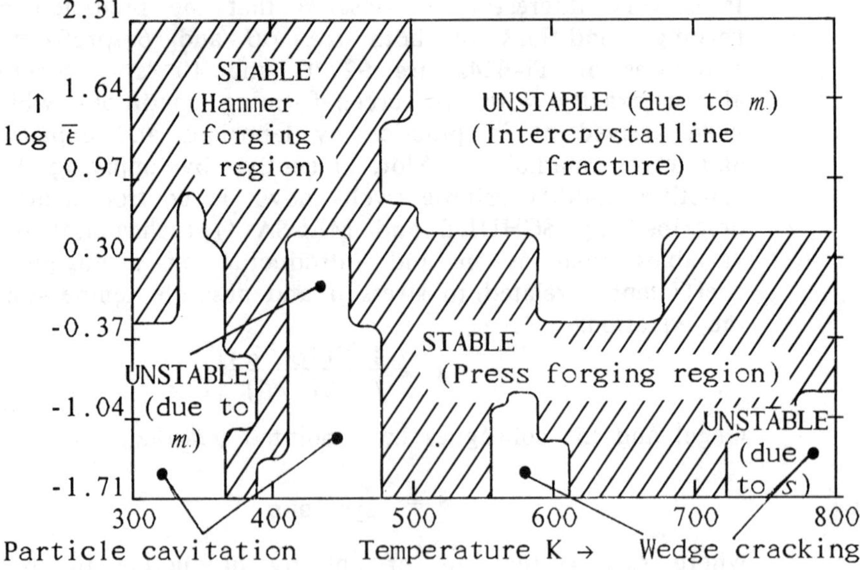

Fig. 1.16 Stability map showing safe processing regions for Al-5Si alloy, at $\bar{\epsilon} = 0.6$ (courtesy of GOPINATH[28])

1.3.4 *Summary and Conclusions*

It seems that the mapping techniques being developed by Gegel and his group offer the most useful approach for developing better mathematical modelling of material behaviour. Data-bases should be set up for several materials of interest, using the same techniques which have given such excellent predictions of actual behaviour for the materials used by Gegel.

At first sight, many researchers find difficulty in accepting the concept of dissipator content and co-content. Also, it has been observed that variables such as T and $\bar{\epsilon}$ are being treated as state variables (much as pressure, volume and temperature in gas thermodynamics), whereas they must be dependent on the history of their development. For example, if a material is taken to a certain value of temperature and total strain without passing through any phase transformation it will then have a different stress-strain-rate relationship than if it __had__ passed through a phase change.

However, the method manifestly works. It is not actually necessary to use the concept of *efficiency*, viz. $\eta = J/J_{max}$. However, it is easier to correlate the parameter η with the microstructures evaluated over various regions of the processing maps and thereby arrive at more meaningful conclusions for understanding the intrinsic *workability* of materials. Similar results could be obtained simply by maximizing the m-value of Equation 1.18 (up to its maximum stable value of unity), i.e. by developing three-dimensional plots and mappings with m as the ordinate rather than $\eta = 2m/(m+1)$, since:-

$$\eta m + \eta = 2m, \therefore m(2-\eta) = \eta, \text{ and } m = \eta/(2-\eta) \qquad (1.25)$$

There is, in fact, almost a linear relationship between η and m for values of η and m between 0 and 1 as shown in Fig. 1.17.

Since a value of $m = 1$ corresponds with Newtonian viscosity, i.e. the viscosity possessed by most simple liquids, this is obviously the most efficient way to deform a material (by making it approximate to the liquid state). The higher the value of m, the easier (more efficient) is the process of deformation going to be (e.g. metals becoming "super-plastic"). However, it is not possible so easily to correlate the value of m with observed or predicted microstructures as is the efficiency η.

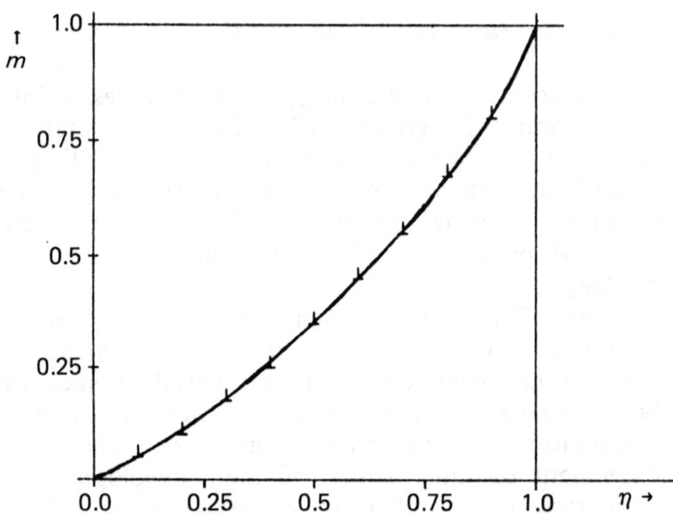

Fig. 1.17 Variation of m with η for Equation 1.25

It may be concluded that the best approach to constitutive equations is that of developing deformation and fracture mechanism maps, on the lines proposed by PRASAD et al.[26]. The *deformation efficiency* concept is attractive and has demonstrably led to a better understanding of the important parameters to be developed to give a meaningful data-base for any given material of interest. The *Lyapunov function stability criteria* seem to provide powerful tools for delineating stable and unstable regions on these mappings of deformation behaviour.   Apparently, these regions so discovered do actually correspond to material instabilities well-known to materials scientists dealing with both metals and non-metals. One remaining problem, however, is that of adequately accounting for the history of the deformation.

Modelling of dynamic material behaviour in hot deformation can most easily be included by using this mapping approach. The data-base so developed could then be used to provide data as a function of temperature, strain, strain rate and time (to include the history of deformation) which can then be fed into comprehensive computer solutions for rolling of the type already developed by ALEXANDER[29], VENTER and ABD-RABBO[30], and HALAL and KAFTANOGLU[31].

However, these computer solutions are all of the *slab* type and do not include either temperature or strain rate effects.   Much work has been done in industry on this problem, but unfortunately it has not been published widely.

1.4 Physical Properties of Importance in Manufacture and Use

Analysis of Hot Forming, Casting or use at high temperatures requires a knowledge of Thermal Conductivity κ, Specific Heat c, and Thermal Expansion α.

The latter is especially important in resistance to thermal shock. A material such as silicon - aluminium - oxy-nitride has a much lower thermal expansion coefficient than, say, aluminium oxide and is much more resistant to cracking when suddenly heated or cooled. Thermal conductivity is also valuable in reducing temperature gradients in a material or composite.

When composites are used it is necessary to consider the compatibility of expansion coefficients, otherwise high stresses and even plastic deformation may occur.

1.5 Chemical Properties

Oxidation is clearly important in hot working operations, though in extrusion, for example, there is very little access of air so oxidation is avoided.

The book edited by BRAITHWATE[32] explains how chemical reaction with lubricants plays a major role in the effectiveness of lubricants. Titanium, for example, is very difficult to lubricate because of its thin inert oxide skin. SCHEY[33] points out that stainless steels present the same problem.

Corrosion is also, of course, a critical feature. A high proportion of all metallic scrap arises from rusting and related electrochemical degradation processes.

Much attention has been given to plating and other protective coatings to avoid corrosion and also to improve other surface properties such as hardness and wear resistance (see SWANN et al.[34], PETERSON and WINER[35]).

1.6 Workability

Although industrial producers will readily recognize some alloys as being easier to work than others, there is no simple test or clearly-defined set of properties.

Ductility is a related property, and it is generally considered that a greater reduction of area to failure in a tensile test indicates better workability. While this is true in general terms, the actual fracture in a tensile test depends upon the complex stress state in the neck region for all but the most brittle alloys.

In the comprehensive book by ATKINS and MAI[21], various other criteria are mentioned which have been proposed on the supposition that fracture in tension depends upon the total plastic work or the tensile plastic work expended in the deformation. It is very difficult to obtain accurate experimental data because the final stress state before fracture is frequently ill-defined. Recent analytical work by CLIFT et al.[36] using finite-element plasticity has given a better insight into these problems.

Hot workability is often described in terms of the number of revolutions to failure in a torsion test as described by HODIERNE[37], but again the analytical significance is obscure. Torsion testing is also a valuable indicator of cold workability, but the results depend strongly upon the axial constraint, possibly because fracture and rewelding occur during the test, so the conditions must be carefully controlled.

REFERENCES

(1)   ALEXANDER, J.M., BREWER, R.C. and ROWE, G.W.,
       Manufacturing Technology: Vol 2. Engineering Processes.
       Ellis Horwood, Chichester, 1986.
(2)   British Standard Specification BS.18, 1956.
(3)   ROLLASON, E.C., *Metallurgy for Engineers.*
       Edward Arnold, London (4th Ed.), 1973.
(4)   BAILEY, A.R., *A Textbook of Metallurgy.*
       Macmillan, London, 1964.
(5)   ROWE, G.W., *Elements of Metalworking Theory.*
       Arnold, London, 1979.
(6)   HOFFMAN, O. and SACHS, G., *Introduction to the
       theory of Plasticity for Engineers.* McGraw-Hill, 1953.
(7)   BRINELL, J.A., "Methods of Testing Steel", *Cong. Int.
       Methodes d'Essai, Paris* 1900,2.
(8)   TABOR, D, *The Hardness of Metals.*
       Oxford University Press, 1951.
(9)   ROWE, G.W., SMETHURST, E. and DOWNING, A.H.,
       "A Reassessment of Accuracy in Diamond Pyramid
       Hardness Testing", *J. Mat. Sci. Letters* 1982,1,109-112.

(10) American Society for Testing Metals, *Standard E 18*.

(11) CADDELL, R.M., *Deformation and Fracture of Solids*. Prentice-Hall, New Jersey, 1980.

(12) CRANE, F.A.A. and CHARLES, J.A., *Selection and Use of Engineering Materials*. Butterworths, 1984.

(13) VAN VLACK, L.H., *Elements of Materials Science*. Addison-Wesley, Reading, Mass:, 1959.

(14) EL-HELIEBY, S.O.A. and ROWE, G.W., "Influences of Surface Roughness and Residual Stress on Fatigue Life of Ground Steel Components", *Metals Technology* 1980,7,221-225.

(15) WAINWRIGHT, S.A., BIGGS, W.D., CURREY, J.D. and GOSLINE, J.M., *Mechanical Design in Organisms*. Arnold, London, 1976.

(16) TIPPER, C.F., "The Brittle Fracture of Metals at Atmospheric and Sub-Zero Temperatures", *Metals Reviews* 1957,2,195-261.

(17) DIETER, G.E., *Mechanical Metallurgy*. McGraw-Hill, New York, 1961.

(18) ANDRADE, E.N.Da C. and CHALMERS, B., "The Resistivity of Polycrystalline Wires in Relation to Plastic Deformation and the Mechanism of Plastic Flow", *Proc. Roy. Soc. London* 1932,**138A**,348-374.

(19) GRIFFITH, A.A., "The Phenomena of Rupture and Flow in Solids", *Phil. Trans. Roy. Soc. London* 1920,**221A**,163-198.

(20) HERTZBERG, R.W., *Deformation and Fracture Mechanics of Engineering Materials*. Wiley, New York, 1976.

(21) ATKINS, A.G. and MAI, Y.W. *Elastic and Plastic Fracture*. Ellis Horwood, Chichester, 1985.

(22) HOSFORD, W.F. and CADDELL, R.M., *Metal Forming: Mechanics and Metallurgy*. Prentice-Hall, New Jersey, 1983.

(23) ALEXANDER, J.M., "On Problems of Plastic Flow of Metals", *Plasticity Today* (Ed. H. Sawczuk), W. Olzak Memorial Volume, Int. Centre for Mechanical Sciences, Udine, Italy, June 1983. To be published by Elsevier, Applied Science Publishers Ltd., 1986.

(24) ASHBY, M.F. and FROST, H.J., "The Kinetics of Inelastic Deformation above $0°K$" *Constitutive Equations in Plasticity* (Ed. A.S. Argon). M.I.T. Press, 1975, p.117.

(25) KELLY, A., *Strong Solids*. Clarendon Press, Oxford,1966.

(26) PRASAD, Y.V.R.K., GEGEL, H.L., DORAIVELU, S.M., MALAS, J.C., MORGAN, J.T., LARK, K.A. and BARKER, D.R., "Modelling of Dynamic Material Behaviour in Hot Deformation: Forging of Ti-6242", *Met.Trans.A.* 1984,15A,1883.

(27) SCHULTZ, D.G. and MELSA, J.L., *State Functions and Linear Control Systems.* McGraw-Hill, 1967,155-195.

(28) GOPINATH, S., *Automation of the Data Analysis System Used in Process Modelling Applications.* M.S. Thesis, Ohio University, Athens, Ohio 45701, 1986.

(29) ALEXANDER, J.M., "Micro Computer Programs for Flat Rolling and Extrolling", *Metal Forming and Impact Mechanics* (Ed. S.R.Reid), William Johnson Commemorative Volume, Pergamon Press, 1986,91.

(30) VENTER, R.D. and ABD-RABBO, A.A., "Modelling of the Rolling Process - I and II", *Int.J.Mech.Sci.* 1980,22,83-92,93-98.

(31) HALAL, A.S. and KAFTANOGLU, B., "Computer-Aided Modelling of Hot and Cold Rolling of Flat Strip", *A.S.M.E. Conf. on Int. Computers in Eng., Las Vegas, Nevada* 1984.

(32) BRAITHWAITE, E.R. (Ed.), *Lubrication and Lubricants.* Elsevier, Amsterdam, 1967.

(33) SCHEY, J.A. (Ed.), *Metal Deformation Processes: Friction and Lubrication.* Marcel Dekker, New York, 1970.

(34) SWANN, P.R., FORD, F.P. and WESTWOOD, A.R.C., *Mechanisms of Environment-Sensitive Cracking of Materials.* Metals Society, 1977.

(35) PETERSON, M.B. and WINER, W.O. (Ed.), *Wear Control Handbook.* A.S.M.E., New York, 1980.

(36) CLIFT, S.E., HARTLEY, P., STURGESS, C.E.N. and ROWE, G.W., "Fracture Initiation in Plane-Strain Forging", *Proc. 25th Int. M.T.D.R. Conf. Macmillan*, 1985,25,413-419.

(37) HODIERNE, F., "A Torsion Test for Use in Metalworking Studies", *J. Inst. Metals* 1963,91,267-273.

Chapter 2

METALLURGICAL CONSIDERATIONS

2.1 Introduction

It would be impossible to cover the whole science of metallurgy in a single chapter of a book of this size. What will be attempted here is to give a thumbnail sketch of some of the main considerations involved. Broadly speaking the subject can be split into two branches – production metallurgy dealing with the melting and formulation of metallic alloys, and physical metallurgy which attempts to explain the observed behaviour of metals in terms of their atomic structure.

The treatment will of necessity be brief and may suffer from the omission of what many metallurgists would consider to be topics of importance. The books by ROLLASON[1], SMALLMAN[2], COTTRELL[3], CADDELL[4], VAN VLACK[5], and PASCOE[6], are typical of the modern approach. It may be simply stated as being concerned with the atomic or submicroscopic physical structure of materials, on the basis of which can be explained many of the characteristic properties of any particular material, be it metal or non-metal.

Interesting though this approach is, it is still necessary for the practising engineer to have a more 'earthy' knowledge of metallurgical theory and practice, especially when he is concerned with processing the new high strength alloys, for example. He needs to know such things as the effect of varying the alloy composition on the hot working range, or the effect of changes in heat treatment procedure. He is less concerned with the detailed mechanism of the interaction of dislocations or solute atoms in causing strain hardening or strain ageing, although it is helpful for him to have an over-all picture of such phenomena. A most useful discussion of the amount of metallurgical knowledge needed by different types of engineer is given by BALL[7], and has been used as a guide in the preparation of this chapter. There is also an interesting book by DIETER[8], who gives a useful account of the mechanical and metallurgical properties of metals, their development, determination, and inter-relation, and another authoritative text by CRANE and CHARLES[9].

2.2 The Solidification of Metals

The process of solidification is generally the first stage in the production of a metal. Quite often the liquid metal is prepared and poured directly into a mould having the desired final shape of the component it is required to make which is then customarily called a casting (as distinct from an ingot). This process, although at first sight attractive, has many disadvantages, both technically from the point of view of the difficulty of producing castings free from defects whilst having the necessary strength* (and toughness), and economically. By far the greatest tonnage of metal produced is cast in the form of large 'ingots' which are subsequently worked by processes such as forging, extrusion and rolling. In this way the original cast structure is broken up to give small grain size and consolidated to give a stronger material in the form of sheet, rod, bar, or sections suitable for manufacturing into engineering components. Thus, it is important to understand the nature of the cast structure of metals and what is the effect on it of plastic deformation.

2.2.1 *The Crystallization Process*

The process of crystallization or freezing of a metal is rather similar to that of freezing a liquid such as water. Considering a pure molten metal, as it cools it gives out heat, and during the solidification process at the melting temperature additional latent heat has to be extracted, so that a typical cooling curve might be as illustrated in Fig. 2.1.

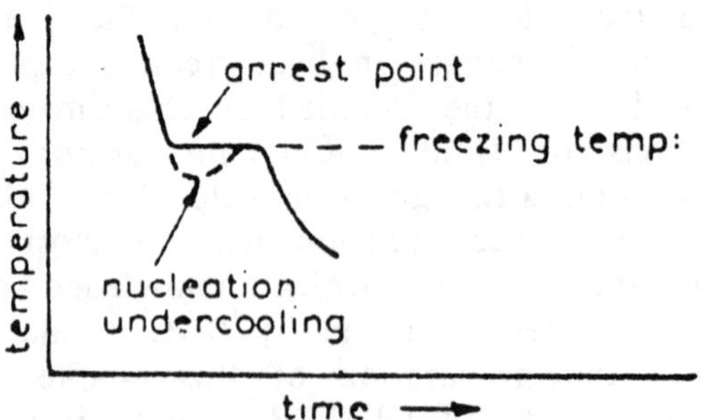

Fig. 2.1 Typical cooling curve

Typical values for the heat required to be supplied to some common metals are shown in Table 2.1.

TABLE 2.1

Metal	Melting temperature °C	Latent heat of fusion J/g	Total heat to raise from room temperature to the molten condition J/g
Aluminium	660	386	1047
Brass	900	167	582
Magnesium	650	198	698
Cast Iron	1220	293	1645

Crystals begin to form at nuclei, and the process of nucleation is assisted if solid foreign particles are present in the liquid metal. If the metal is very pure, nucleation may be difficult, requiring a considerable drop in temperature below the real freezing-point, as illustrated in Fig. 2.1 by the broken line, before crystallization occurs. A similar effect is the familiar "supercooling" of water.

The actual mechanism by which crystals form from the molten metal during freezing is complex and leads in the case of impure metals and alloys to the formation of a well-known fir-tree type of crystal known as a *dendrite,* illustrated in Fig. 2.2. Freezing proceeds from crystal corners so that fingerlike extensions appear which protrude into cooler regions of the melt where the growth is accelerated, additional arms forming as illustrated in the figure.

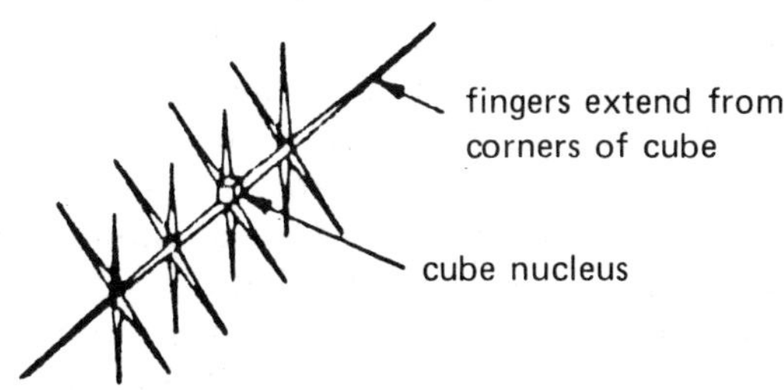

Fig. 2.2 Illustrating formation of a dendrite

Dendrites will form and grow from each nucleus, until contact is made with neighbouring dendrites, the contacting face forming part of the eventual boundary between crystals in the metal. The arms of the dendrites gradually thicken during the freezing process until finally only a solid crystal remains with little trace of the dendritic structure which existed during cooling.   Frequently, impurities are pushed by the growing crystal to the grain boundaries but are trapped between dendrites, revealing their structure in the final metal.

It is clear from this discussion that the final cast structure will depend markedly on the mechanism of formation of dendrites; for example, the more nuclei there are the finer will be the resulting grain size.   The crystal structure is, of course, constant in its direction of orientation in any one dendrite, and hence in any one grain of resulting metal; the orientation is usually different between neighbouring crystals. The final structure is influenced in practice by many factors such as casting temperature, mass of metal, thermal conductivity of mould material and, of course, the particular alloy being cast.   These questions are discussed in more detail later.

2.2.2 *The Atomic Lattice*

A brief discussion of the structure of the atom is given at the beginning of the next chapter, mainly as it affects the types of bonding which can occur between elements.   As discussed in that chapter the atom consists of a nucleus comprising heavy positively charged protons and electrically neutral neutrons, around which revolve a number of light negatively charged electrons.   The electrons revolve in groups of stable orbits called shells each of which can contain only a certain maximum number of electrons, e.g. 2 in the first shell (K shell), 8 in the second shell (L shell), 18 in the third shell (M shell), etc.   It is found that the chemical properties of the elements are closely related to the number of valency electrons in the outer shells, and the elements may be arranged in a *periodic table* in which groups of elements have similar properties.   The periodic table is shown in Table 2.2 on p. 34, and on p. 35 is an alphabetical list of the names of the elements, their symbols and atomic numbers (the table was originally proposed for the then known elements by MENDELEEV in 1869).   The idea that an atom comprises a central positively-charged nucleus surrounded by negative electrons was proposed by RUTHERFORD in 1911, and the quantum model of the atom with its electron shells was developed by BOHR in 1913.

TABLE 2.2 *The periodic table of the elements*

(Based on Appendix I of *Metallurgy for Engineers*, by E. C. Rollason, including data abstracted from the Year Book of the Institution of Metallurgists, 1961–1962.)

METALS ———————————————— | NON-METALS ————

Period	I	II				GROUPS					I	II	III	IV	V	VI	VII	0
Electropositive valency	1	2									1	2	3	4				
Electronegative valency														4	3	2	1	0
1	1 H 1·008																	2 He 4·003
2	3 Li 6·940	4 Be 9·013									Li	Be	5 B 10·82	6 C 12·01	7 N 14·008	8 O 16	9 F 19·00	10 Ne 20·183
3	11 Na 22·997	12 Mg 24·32				*Metals with variable valency, i.e. transition groups*					Na	Mg	13 Al 26·97	14 Si 28·09	15 P 30·974	16 S 32·066	17 Cl 35·457	18 A 39·944
4	19 K 39·100	20 Ca 40·08	21 Sc 44·96	22 Ti 47·90	23 V 50·95	24 Cr 52·01	25 Mn 54·93	26 Fe 55·85	27 Co 58·94	28 Ni 58·69	29 Cu 63·54	30 Zn 65·37	31 Ga 69·72	32 Ge 72·60	33 As 74·91	34 Se 78·96	35 Br 79·916	36 Kr 83·80
5	37 Rb 85·48	38 Sr 87·63	39 Y 88·92	40 Zr 91·22	41 Nb 92·91	42 Mo 95·95	43 Tc (98·91)	44 Ru 101·7	45 Rh 102·91	46 Pd 106·7	47 Ag 107·873	48 Cd 112·41	49 In 114·76	50 Sn 118·70	51 Sb 121·76	52 Te 127·61	53 I 126·91	54 Xe 131·3
6	55 Cs 132·91	56 Ba 137·36	57 La 138·92 †	72 Hf 178·6	73 Ta 180·88	74 W 183·92	75 Re 186·31	76 Os 190·2	77 Ir 193·1	78 Pt 195·23	79 Au 197·2	80 Hg 200·61	81 Tl 204·39	82 Pb 207·21	83 Bi 209·00	84 Po (210)	85 At (210)	86 Rn 222
7	87 Fr (223)	88 Ra 226·05	89 Ac 227 ‡	(104)	(105)	(106)	(107)	(108)										

† Lanthanide or rare earth series.

57 La 138·92	58 Ce 140·13	59 Pr 140·92	60 Nd 144·27	61 Pm (145)	62 Sm 150·43	63 Eu 152·0	64 Gd 156·9	65 Tb 159·2	66 Dy 162·46	67 Ho 164·94	68 Fr 167·2	69 Tm 169·4	70 Yb 173·04	71 Lu 174·99

‡ Actinide or transuranic series.

89 Ac 227	90 Th 232·12	91 Pa 231	92 U 238·07	93 Np 237	94 Pu (239)	95 Am (243)	96 Cm (248)	97 Bk (249)	98 Cf (251)	99 Es (254)	100 Fm (255)	101 Mv (256)	102 No (256)	(103)

Non-metals are near to right-hand corner. Metals near to line parallel dividing metals from non-metals (Al Zn Sn Pb) have oxides which can behave as acids or bases, i.e. show some of the properties of non-metals. Elements in the same vertical column (i.e. the same valency electrons) have similar properties, e.g. Be, Mg, Zn, Cd; or Cr, Mo, W; or all the rare earths and transuranic elements. The atomic number is shown above each symbol, atomic weight below, brackets indicating approximate values.

Table of the elements in alphabetical order

Element	Symbol	Atomic number	Atomic weight
Actinium	Ac	89	227
Aluminium	Al	13	26·97
Americium	Am	95	(243)
Antimony	Ab	51	121·76
Argon	A	18	39·944
Arsenic	As	33	74·91
Astatine	At	85	(210)
Barium	Ba	56	137·36
Berkelium	Bk	97	(249)
Beryllium	Be	4	9·013
Bismuth	Bi	83	209·00
Boron	B	5	10·82
Bromine	Br	35	79·916
Cadmium	Cd	48	112·41
Caesium	Cs	55	132·91
Calcium	Ca	20	40·08
Californium	Cf	98	(251)
Carbon	C	6	12·010
Graphite		-	-
Diamond			
Cerium	Ce	58	140·13
Chlorine	Cl	17	35·457
Chromium	Cr	24	52·01
Cobalt	Co	27	58·94
Copper	Cu	29	63·54
Curium	Cm	96	(248)
Dysprosium	Dy	66	162·46
Einsteinium	Es	99	(254)
Erbium	Er	68	167·2

Element	Symbol	Atomic number	Atomic weight
Europium	Eu	63	152·0
Fermium	Fm	100	(255)
Fluorine	F	9	19·00
Francium	Fr	87	(223)
Gadolinium	Gd	64	156·9
Gallium	Ga	31	69·72
Germanium	Ge	32	72·60
Gold	Au	79	197·2
Hafnium	Hf	72	178·6
Helium	He	2	4·003
Holmium	Ho	67	164·94
Hydrogen	H	1	1·008
Indium	In	49	114·76
Iodine	I	53	126·91
Iridium	Ir	77	193·1
Iron	Fe	26	55·85
Krypton	Kr	36	83·80
Lanthanum	La	57	138·92
Lead	Pb	82	207·21
Lithium	Li	3	6·940
Lutetium	Lu	71	174·99
Magnesium	Mg	12	24·32
Manganese	Mn	25	54·93
Mendelevium	M	101	(256)
Mercury	Hg	80	200·61
Molybdenum	Mo	42	95·95

Element	Symbol	Atomic number	Atomic weight
Neodymium	Nd	60	144·27
Neon	Ne	10	20·183
Neptunium	Np	93	237
Nickel	Ni	28	58·69
Niobium	Nb	41	92·91
Nitrogen	N	7	14·008
Nobelium	No	102	(256)
Osmium	Os	76	190·2
Palladium	Pd	46	106·7
Phosphorus	P	15	30·974
Black		--	--
Red		--	--
White		--	--
Platinum	Pt	78	195·23
Plutonium	Pu	94	(239)
Polonium	Po	84	(210)
Potassium	K	19	39·100
Praseodymium	Pr	59	140·92
Promethium	Pm	61	(145)
Protactinium	Pa	91	231
Radium	Ra	88	226·05
Radon	Rn	86	222
Rhenium	Re	75	186·31
Rhodium	Rh	45	102·91
Rubidium	Rb	37	85·48
Ruthenium	Ru	44	101·7
Samarium	Sm	62	150·43
Scandium	Sc	21	44·96

Element	Symbol	Atomic number	Atomic weight
Selenium	Se	34	78.96
Silicon	Si	14	28·09
Silver	Ag	47	107·873
Sodium	Na	11	22·997
Strontium	Sr	38	87·63
Sulphur	S	16	32·066
Tantalum	Ta	73	180·88
Technetium	Tc	43	(98·91)
Tellurium	Te	52	127·61
Terbium	Tb	65	159·2
Thallium	Tl	81	204·39
Thorium	Th	90	232·12
Thulium	Tm	69	169·4
Tin	Sn	--	118·7
White		50	--
Grey		--	--
Titanium	Ti	22	47·90
Tungsten	W	74	183·92
Uranium	U	92	238·07
Vanadium	V	23	50·95
Xenon	Xe	54	131·3
Ytterbium	Yb	70	173·04
Yttrium	Y	39	88·92
Zinc	Zn	30	65·37
Zirconium	Zr	40	91·22

The main primary bonds which bind together atoms are the ionic bond in which valency electrons are exchanged, and the covalent bond in which valency electrons are shared between atoms. The metallic bond may be regarded as a special case of covalent bonding although this is an over-simplification. In most metals there are only a few valency electrons, which are easily removed leaving atoms or ions comprising a nucleus to which the remaining electrons are firmly held. Each of these ions is positively charged, so that the atomic lattice of the metal is formed from a structure of positive ions and free electrons which move about within the structure to form a negatively charged electron 'cloud' or 'gas'. The forces which bond the atoms together are provided by the attractive forces that act between the positive ions and the electron cloud, which cause the atoms to pack together very closely. The free electrons give the metal its characteristically high electrical and thermal conductivity, and result in the atoms forming a 'close-packed' array to give the crystal lattice characteristic of metals. The electron cloud also absorbs light energy so that all metals are opaque to transmitted light.

Atoms which are attracted towards one another by the attractive forces due to interchanging or sharing valency electrons eventually reach an equilibrium position in which these attractive forces are just balanced by the repulsive forces due to the like polarities of the nuclei. This equilibrium state can then be disturbed by various means; the addition of thermal energy by raising the temperature will produce motion of the atoms, eventually separating the atoms completely, so that the metal will become liquid and even gaseous if the temperature is raised enough. The bonds between atoms of a metal are strong but the atoms are able to alter their relative positions by dislocation motion, to be discussed later, which is the reason for the characteristic plasticity of metals. Either electrical or mechanical energy can be supplied for this purpose, deforming and even fracturing the material. The repulsive forces between nuclei become extremely great as the distance between them lessens, so that all materials are extremely resistant to all-round compressive forces and will return elastically to their former dimensions on release of the pressure.

The size of an atom in such a metallic structure may be regarded as being described by its diameter, which is equal to the minimum equilibrium distance between the centres of neighbouring atoms. In iron, for example, this diameter is 2.476 Å (Angström units, 1 Å = 10^{-7} mm or 10^{-4} micrometres).   As has been seen the equilibrium distance can be affected in various ways, one of the most important being due to the crystalline array of the atomic lattice.   Most solid metals form atomic patterns based on either a cubic or hexagonal arrangement, and these symmetrical arrangements generally lead to high ductility or plasticity.   Arrangements of lower symmetry such as occur in bismuth and antimony lead to brittleness of the metal.

When in the gaseous form, the atoms are completely free and move about haphazardly.   In the liquid state the atoms are still without any regular arrangement but are much closer together than in a gas.   If two or more of the different metals are mixed together in the liquid state then the atoms of the different metals will be randomly distributed throughout the volume of the liquid which will possess properties different from any of its constituents.   When this molten alloy solidifies on cooling, the atoms must take up a definite crystalline array, and there are a variety of ways in which this can occur. If the metals are insoluble in the solid state the different atoms form the kinds of crystal lattice associated with each one singly.

In many cases the solidification of an alloy of two metals results in the formation of one kind of crystal in which atoms of both metals are present.   These are (a) *solid solutions*, and (b) *intermediate constituents*, sometimes two of the former, sometimes one of each.   Solid solutions occur most readily when the two metallic constituents have atoms of similar size and lattice arrangement and equal numbers of valency electrons.   This will occur if the elements are in the same vertical column (or *Group*) in the periodic table (Table 2.2). If the elements are in different groups the formation of intermediate constituents is more likely than solid solutions. Intermediate constituents (or chemical compounds) have a crystal structure different from either of the elements.   As might be expected, the lattice arrangement of the solid solution is a slightly distorted version of that of one of the constituents.   Solid solutions are similar to the more familiar liquid solutions in which one of the elements is dissolved in the solvent so that its identity is lost (e.g. sugar in water). *Brass* containing up to about 39 per cent of zinc is a solid solution of copper and zinc, and has a face-centred cubic structure like pure copper, and some of the copper atoms are replaced randomly by zinc atoms.   (Zinc has a hexagonal close-packed structure.)   Such a solid solution, which is quite common, is called a *substitutional solid solution*.   The amount of zinc which will go into substitutional solid solution with copper to form brass is limited, however, to about 39 per cent; there is no such limitation with nickel, which will form a substitutional solid solution in any proportion with copper. Similarly, there is a limit to the amount of tin which will go into solid solution with copper to form tin-bronze; tin in excess of this *solubility limit* must form another phase.

The important criterion in specifying solubility limits is the number of substituted atoms rather than their weight. Since engineers generally express composition as percentage by weight it is necessary to know how to express weight per cent in terms of atomic per cent and vice versa.   An electron weighs only about 0.005 as much as a neutron or proton so that the weight of an atom is nearly proportional to the total weight of protons and neutrons in the nucleus, which is called the *atomic weight*.   Atomic weights range from 1.008 for hydrogen to nearly 250 for some of the transuranic elements. Atomic weight is expressed in grams per gram-atom, a gram-atom comprising 6.02×10^{23} *atoms (Avogadro's number)*. Thus the weight of one atom is given by the equation:-

$$\text{Weight of atom} = \frac{\text{Atomic Weight}}{6.02 \times 10^{23}} \quad \text{(grams)} \qquad (2.1)$$

Consider for example a solid solution of 100 grams of brass made up of 70 grams of copper and 30 grams of zinc; what will be the atomic per cent of such an alloy?   The atomic weight of copper is 63.54 and the atomic weight of zinc is 65.37.   Therefore the 70 grams of copper will contain:-

$$\frac{70}{63.54} \times 6.02 \times 10^{23} \text{ atoms} = 6.6320 \times 10^{23} \text{ atoms},$$

whilst the 30 grams of zinc will contain:-

$$\frac{30}{65.37} \times 6.02 \times 10^{23} = 2.7627 \times 10^{23} \text{ atoms}.$$

The total number of atoms is 9.394×10^{23}, so that the atomic per cent of copper is 70.593 a/o, and of zinc 29.407 a/o.

The atomic weight should not be confused with the *atomic number*, which equals the number of protons in the nucleus, and therefore the number of electrons surrounding the nucleus of a neutral atom.   Thus the atomic number of helium which has two protons and two neutrons in its nucleus is 2, whilst its atomic weight is 4.003.   The atomic weight of all the elements is shown in the periodic table (Table 2.2).

In certain alloys the atoms of the second metal occupy

preferential sites in the lattice of the parent metal, forming a regular array in conjunction with the parent lattice, called a *super-lattice*. A typical example of this is the β-phase in brass, which above about 460°C has a random structure, but below has an arrangement in which, if the structure is regarded as an interpenetration of two simple cubic lattices, the copper atoms exist at the corners of one and the zinc atoms at the corners of the other.

An *interstitial solid solution* is formed when the atoms of the second element occupy spaces between those of the primary metal instead of replacing them. The elements which will form this type of solid solution generally have atoms somewhat smaller than those of the parent metal, for example, carbon (0.77 Å radius) in γ-iron (approx. 1.26 Å radius) which forms the basis of steels. Hydrogen, boron, and nitrogen also form interstitial solid solutions in metals, their atoms being small (0.46, 0.97, and 0.71 Å radius respectively in their equilibrium states).

2.2.3 *The Crystal Structure*

There are six main patterns which are possible in the formation of crystals. The first five of these may be described by Fig. 2.3, in which the circles are intended to represent the centres of adjacent atoms and the x,y,z axes are not necessarily orthogonal. The distance between the atoms is determined by their associated electrical fields, for example, in the simple cubic structure in which $a=b=c$, the distance a would be equal to twice the radius of the atom. The crystal structure is made up of a three-dimensional repeated array, of which the four atoms illustrated in Fig. 2.3 may be regarded as a unit cell. The possible patterns are as follows:-

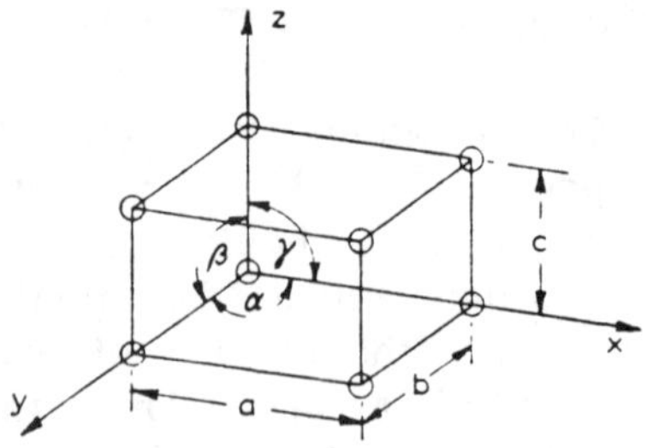

Fig. 2.3 The basic crystal lattice, illustrating rotation

(1) Cubic: $a=b=c$; $\alpha=\beta=\gamma=90\,^\circ$.
(2) Rhombohedral: $a=b=c$; $\alpha=\beta=\gamma\neq90\,^\circ$.
(3) Tetragonal: $a=b\neq c$; $\alpha=\beta=\gamma=90\,^\circ$.
(4) Orthorhombic: $a\neq b\neq c$; $\alpha=\beta=\gamma=90\,^\circ$.
(5) Monoclinic: $a\neq b\neq c$; $\alpha=\gamma=90\,^\circ$, $\beta\neq90\,^\circ$.
(6) Triclinic: $a\neq b\neq c$; $\alpha\neq\beta\neq\gamma$ (there are no right angles).
(7) Hexagonal: $a=b\neq c$; $\beta=\gamma=90\,^\circ$, $\alpha=120\,^\circ$.

The seventh (hexagonal) structure is illustrated in Fig. 2.4, which is sometimes regarded as being made up of three smaller *rhombic* cells of the type shown on the right of Fig. 2.4.

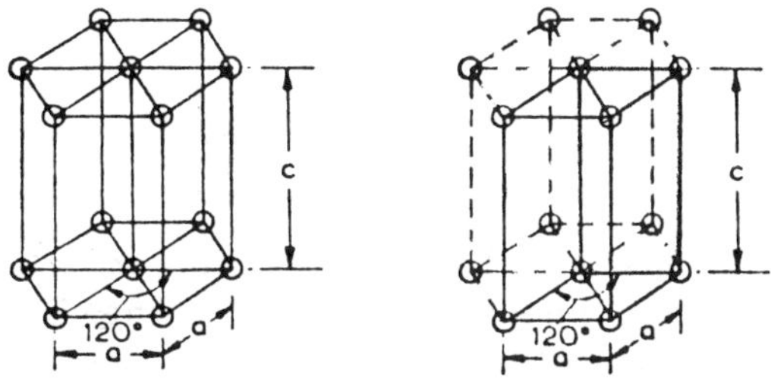

Fig. 2.4 The hexagonal lattice

The simple geometrical arrangements described above may be combined in over 200 different ways to give different lattice structures, but only the more important of these will be discussed here.

Iron has a *body-centred cubic* structure (b.c.c.), as illustrated in Fig. 2.5. Each atom may be regarded as being either at the centre or the corner of a unit cube in the cubic lattice which is extended typically as indicated by the broken lines.   Thus each atom is surrounded by an identical configuration of other atoms.   Although a circle has been drawn in Fig. 2.5 to indicate the location of each atomic centre, the distances between atoms are determined by their spheres. Since the atomic diameter or size may conveniently be regarded as the equilibrium distance between centres, it is possible to interrelate crystal lattice dimensions such as a in Fig. 2.5 with the atomic radius r, by simple geometry.   The distance a between adjacent corners of the unit cell is called the *lattice constant* or *lattice parameter*.   This is illustrated in Fig. 2.5 for the b.c.c. structure, by considering a diagonal of the unit cube. From the theorem of Pythagoras:-

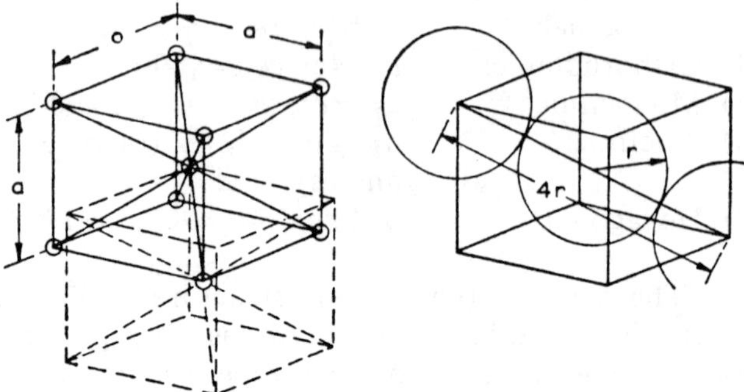

Fig 2.5 The body-centred cubic (b.c.c.) lattice

$$a^2 + a^2 + a^2 = 16r^2$$

$$\therefore \quad a = \frac{4r}{\sqrt{3}} \qquad\qquad (2.2)$$

Hence, since $r = 1.238$ Å for iron, $a = 2.86$ Å.

Other important metals having b.c.c. structure are, for example, chromium, vanadium, tungsten, molybdenum, and tantalum.

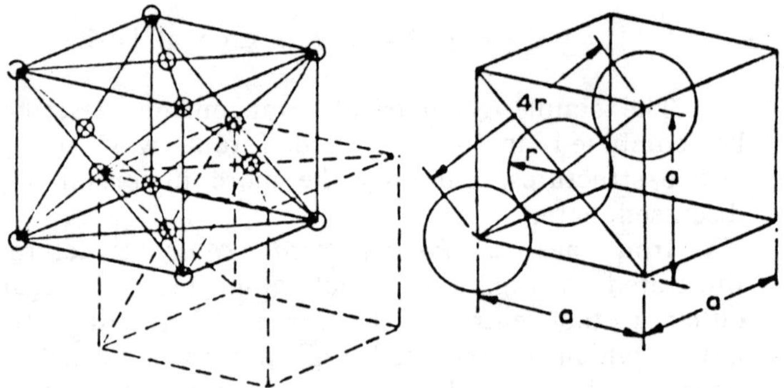

Fig. 2.6 The face-centred cubic (f.c.c.) lattice

Copper, aluminium, lead, nickel, gold, and silver also possess a cubic lattice, but instead of there being an atom at the centre of each cube, there is one at the centre of each face of the cube, as illustrated in Fig. 2.6. The structure is extended typically as shown by the broken line, from which we see that each atom may be regarded as being either at the corner or the face centre of a cube. The relation between the lattice parameter a and atomic radius is seen to be:-

$$a = \sqrt{8}\,r \qquad (2.3)$$

This *face-centred cubic* structure (f.c.c.) is very common in metals and is also found in non-metals, in which atoms of different elements may form the lattice in a symmetrical arrangement.

The most important example of the hexagonal structure illustrated in Fig. 2.4 is that of graphite. In metals the structure is usually more dense, being referred to as *hexagonal close-packed* (h.c.p.), as illustrated in Fig. 2.7.

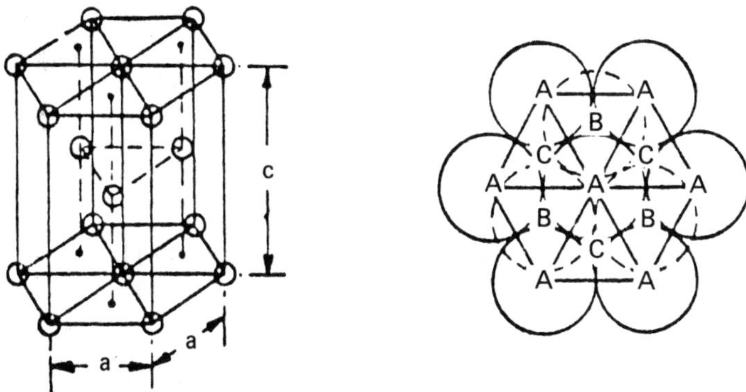

Fig. 2.7 The hexagonal close-packed (h.c.p.) lattice

It is characterized by each atom being located centrally above and below the space between three atoms in the adjacent layers. Thus each atom contacts six atoms in its own plane and three in each layer above and below making a total of twelve surrounding atoms. It is easily shown that:-

$$a = 2r \quad \text{and} \quad c = 4\sqrt{\frac{2}{3}}\,r \qquad (2.4)$$

The hexagonal close-packed and face-centred cubic structures both represent the most closely-packed arrangement of atoms possible in the lattice structure. Considering the arrangement of atoms shown in the inset diagram in Fig. 2.7, if the letters A represent the centres of one plane of close-packed atoms and the letters B and C represent possible centres of adjacent close-packed planes of atoms below the first plane then the arrangement ABCABCABC..... gives a face-centred cubic lattice whereas the arrangement ABAB..... or ACAC..... gives a hexagonal close-packed structure. A list of some of the more common metals, together with their crystal structures and atomic radii is given in Table 2.3.

TABLE 2.3

Body-centred cubic	Face-centred cubic	Hexagonal close-packed
Chromium (1·246)	Aluminium (1·43)	Beryllium (1·125)
Iron (α and δ) (1·238)	Cobalt (β)(−)	Cadmium (1·486)
Manganese (δ) (−)	Copper (1·275)	Cobalt (α) (1·25)
Molybdenum (1·36)	Gold (1·435)	Magnesium (1·6)
		Titanium (α) 00 (1·46)
Niobium (1·43)	Iridium (1·35)	Zinc (1·33)
Tantalum (1·425)	Iron (γ) (1·26)	Zirconium (α) (1·59)
Titanium (β)(1·43)	Lead (1·745)	
Tungsten (1·41)	Nickel (1·25)	
Uranium (γ) (−)	Palladium (1·37)	
Vanadium (1·313)	Platinum (1·38)	
Zirconium(B)(1·57)	Rhodium (1·342)	
	Silver (1·44)	

Figures in brackets are the atomic radii (shortest distance) in Ångström units, where known (N.P.L.). (Data abstracted from the 1961-62 Year Book, Institution of Metallurgists.)

Lattice parameters and crystal orientation may be observed experimentally by employing X-rays, which can be arranged to be diffracted by the planes of atoms within the crystal. One of the most commonly used methods is that of back reflection in which a monochromatic beam of X-rays is directed on to the specimen. The X-rays are reflected from suitably orientated planes of atoms, on to a photographic plate or film, forming rings whose definition depends both on the crystal size and on the amount of cold work which has been expended on the metal. The determination of lattice parameters is usually carried out by a diffraction method and the density of the lines obtained from this technique depends on the coincidence or otherwise of X-rays diffracted by different planes of atoms. This 'density' depends on the wavelength λ and angle of incidence θ of the X-rays, and the interplanar spacing d of the atoms. The dependence may be calculated from the geometry of the diffracted waves, and is expressed by the Bragg equation:-

$$n\lambda = 2d \sin \theta \qquad (2.5)$$

where n is a whole number representing the 'order' or number of waves occurring in a certain distance within the crystal lattice. In practice λ is known, θ can be measured, n can be inferred and hence the lattice parameter d can be determined.

The usual method of identifying the various atomic planes that exist in a crystal is by using Miller indices which are the smallest integral reciprocals of the intercepts of the plane on the crystal axes x,y,z (Fig. 2.3), in terms of the lattice parameter as unit dimension. Consider the faces of the cube shown in Fig. 2.8.

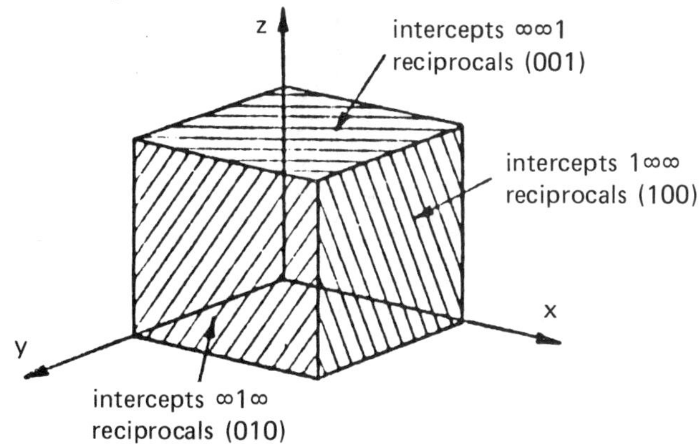

Fig. 2.8 The Miller indices

It will be clear from this figure that the plane normal to the z axis is the (001) plane and the abbreviation is used of replacing '0,0,1 plane' by '(001)', where the ordinary brackets denote 'plane'. Fig. 2.9 denotes other important planes in the cubic lattice — since the actual position of a plane in relation to any origin of x,y,z axes is immaterial it is evident that planes ($\bar{1}11$) and ($1\bar{1}1$) are equivalent (the symbol $\bar{1}$ meaning -1). Similarly, plane ($2\bar{2}2$) is equivalent to plane ($1\bar{1}1$) and the last definition (having the smallest integral numbers) would be used in practice. Direction in the atomic lattice is represented by giving the co-ordinates of the end of a space vector in terms of the lattice parameter, within *square* brackets. Thus the diagonal of the cubic lattice bisecting the positive quadrant is represented by the symbol [111], and other directions are illustrated in Fig. 2.10.

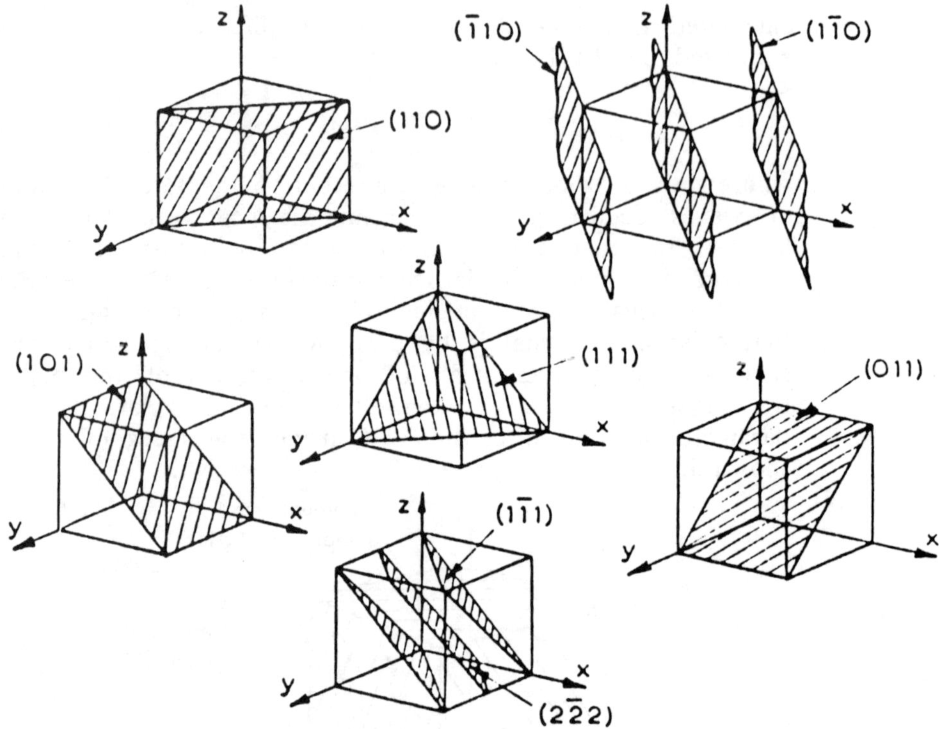

Fig. 2.9 Important planes in the cubic lattice

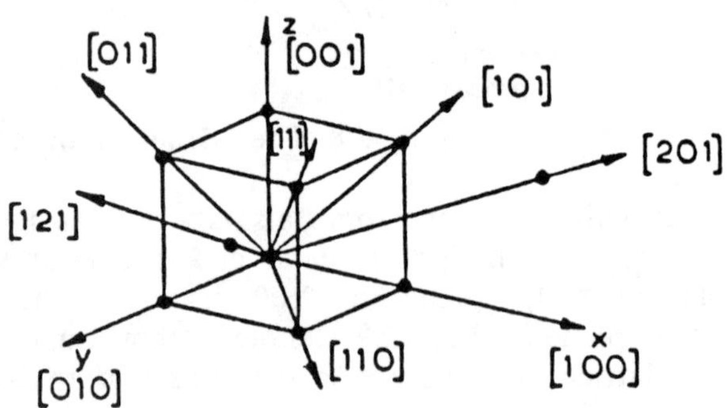

Fig. 2.10 Important directions in the cubic lattice

Pole figures are frequently used in the literature to represent the orientation of crystallographic planes within the material. The material, for example the unit cubic crystal illustrated in Fig. 2.11(a), is imagined to be at the centre of a sphere. A normal to any plane considered will cut the sphere at its pole, e.g. P for the (101) plane in Fig. 2.11(a).

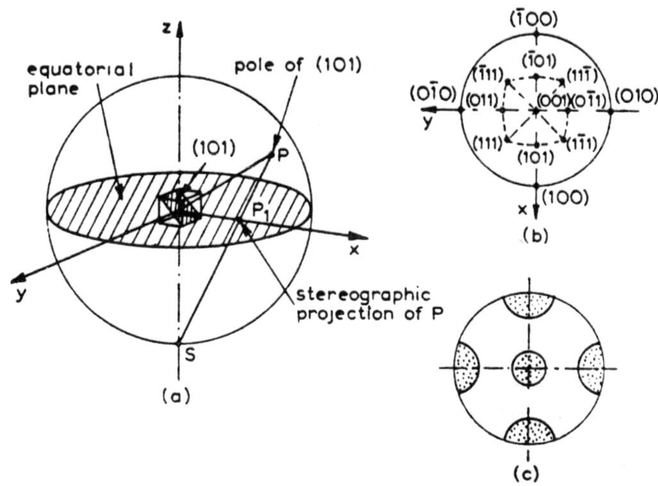

Fig. 2.11 The pole figure

(a) Definitions
(b) The equatorial plane
(c) Typical appearance of pole figure for a metal having preferred orientation

If a line is now drawn from the pole P to the south pole of the sphere S, it will cut the *equatorial plane* of the sphere at the point P_1 which is the *stereographic projection* of the pole P. The stereographic projections of the poles of all the faces and diagonal planes of the unit cube on the equatorial plane will be as shown in Fig. 2.11(b).

A typical pole figure from a polycrystalline metal in which the f.c.c. crystals have preferred orientation is illustrated in Fig. 2.11(c). The metal has been rolled and annealed so that the faces of the crystal cubes (the (100) planes) are aligned so as to be parallel with the rolling direction (so-called 'cube-texture'), and the intensity of the spots in each of the relevant areas of the pole figure reveals the extent of preferred orientation. A metal with completely random orientation of the crystals will give a pole figure in which the spots are distributed evenly all over the field.

·In the polycrystalline metal it is often necessary to give some idea of the size of individual crystals or grains. Obviously the best way would be to determine and quote the number of grains per unit volume of the metal, but this is difficult to achieve in practice. It is customary to quote the number of grains visible per unit area of a prepared section of the metal and quote the grain size as the number of grains per sq. cm. (or per sq. in.).   The American Society for Testing Materials (A.S.T.M.) define grain size N in the following way:-

$$n = 2^{N-1} \qquad (2.6)$$

where n is the number of grains per 10^{-4} in^2 (i.e. observed area of 1 in^2 at $\times$ 100 magnification)

$$\therefore N = 1 + \frac{\log_{10} n}{0.3010} \qquad (2.6a)$$

It is only necessary to quote N to the nearest whole number, so that if $n = 62$, N would be 7 (actually 6.95); if $n = 620$, N would be 10 (actually 10.3); if $n = 6200$, N would be 14 (actually 13.6).

2.2.4 *Equilibrium Diagrams*

The equilibrium behaviour of any pure metal or alloy during heating or cooling can be represented by *thermal equilibrium diagrams* (*phase diagrams* or *constitution diagrams*). Although such diagrams are undoubtedly of limited use since they refer to <u>final</u> equilibrium states which may take some time to achieve in practice, they give a guide to the expected behaviour of the given metal.   There are three cases of importance:

(1) *Pure metals,*
(2) *Primary solid solutions,* in which the crystal structure of the alloy is similar to that of one of the constituents,
(3) *Intermediate constituents,* in which the resulting crystal structure is different from either of the constituents — often referred to as *intermetallic compounds* if both constituents are metallic.

Such a diagram indicates the different types of crystal present at equilibrium and their relative quantities in a particular alloy at a specified temperature.   It will not indicate the spatial disposition of the phases (morphology); for instance, whether they exist in the form of plates or laminae, globules, or films; neither will it give any indication of the rate at which reactions occur.

One of the ways in which equilibrium diagrams are constructed is from cooling curves, and this is best illustrated by an example.   Consider two metals A and B which although mutually and completely soluble in the liquid state are completely insoluble in each other in the solid state. It is important to remember at this stage what is implied by the term solubility. It implies that atoms of one constituent may either replace atoms of the other, or fit into interstices in the lattice structure - if this is not possible then the substances are insoluble in one another. The difference between a mixture, in which two distinct phases are present, and a solution in which only one phase (or structural pattern) occurs, is obvious.

In Fig. 2.12 are illustrated cooling curves (temperature versus time) for the two metals A and B, in different concentrations in the liquid state, their melting point temperatures being T_A and T_B respectively. The cooling curve for the 100 per cent A solution needs no comment, the arrest point at temperature T_A corresponding with the liberation of the latent heat energy during solidification; the cooling curve for the 100 per cent B solution is similar. The metal is solid immediately below the relevant melting temperature for each of these extreme cases. Considering the liquid solution comprising 90 per cent A, 10 per cent B, it is found that during cooling there is now no arrest point at the temperature T_A, due to the addition of the metal B, followed by a short arrest at the temperature T_E which is unrelated to either T_A or T_B, and is lower than both. Solidification does not occur at the single temperature T_A or T_B as was found with either pure metal, but takes place continuously during cooling over the range from T_1 to T_E.

As the percentage of metal B in the liquid solution is increased, to 20 per cent say, this phenomenon again occurs, the first arrest being at an even lower temperature T_2, and shorter than for the 10 per cent solution. The second arrest again occurs at T_E, for a longer period than for the 10 per cent solution. This behaviour continues as the percentage content of B is increased, until a solution is reached at which the arrest points coincide, at temperature T_E.

This particular solution is termed the *eutectic*, the temperature T_E being the *eutectic temperature.* For the example shown it has been assumed that the eutectic composition is 60 per cent A, 40 per cent B. A solution of this composition will solidify immediately at the temperature T_E into a mechanical mixture of the metals A and B. A liquid solution of 80 per cent A, 20 per cent B will start to solidify on cooling, at temperature T_2, dendrites of the pure metal A forming within the liquid which will consequently have a reduced amount of *A* in solution.

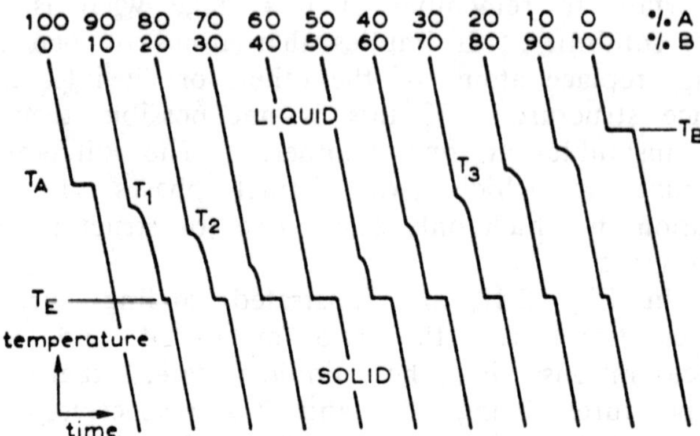

Fig. 2.12 Cooling curves for differing
concentrations of metals A and B

On further cooling liquid *A* will solidify into dendrites at
the temperature T_E.   Thus the final solid will comprise
dendrites of metal *A* surrounded by the eutectic mixture of
metals *A* and *B*.   Similarly a liquid solution of say 30 per
cent *A* and 70 per cent *B* will start solidifying at temperature
T_3 (Fig. 2.12), dendrites of pure metal *B* being formed in the
surrounding liquid of gradually decreasing *B* content.   In this
case, the final solid comprises dendrites of metal *B* surrounded
by the eutectic.   The equilibrium diagram is constructed from
the cooling curves by joining together the traces of the points
as shown for the alloy just considered, in Fig. 2.13.

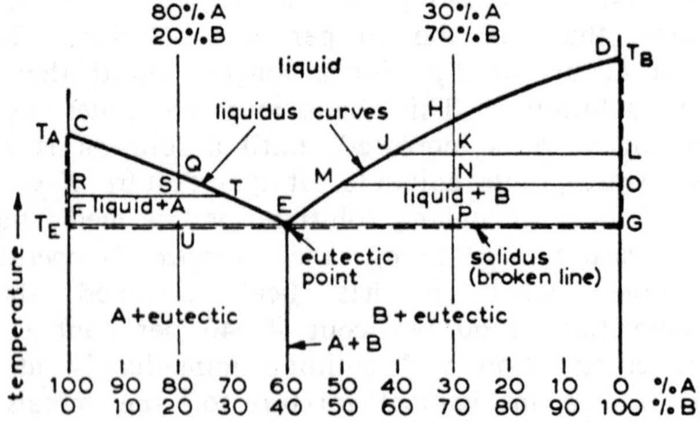

Fig. 2.13 The equilibrium diagram

Considering the 80 per cent A, 20 per cent B solution, during cooling from Q to U the liquid becomes enriched in metal B and in fact follows the curve QTE.   At the temperature RST the (steady-state) alloy must consist of pure metal A dendrites plus liquid of composition T.   Since RS represents the proportion of B in the original liquid and RT = RS + ST represents the proportion of B in the original liquid of composition T, and these must be equal in weight, the following relationship applies:-

$$(RS + ST)L_T = (RS)(A + L_T)$$

where L_T is the weight of the liquid of composition T and A is the weight of the pure metal A present at temperature RST. Hence it is easily shown that:-

$$\frac{A}{L_T} = \frac{ST}{RS} \tag{2.7}$$

At temperature T_E (or FUE), therefore, the ratio of the weight of the pure metal A dendrites to the weight of the liquid eutectic will be given by the dimensions:-

$$\frac{A}{L_E} = \frac{UE}{FU} \tag{2.7a}$$

For this particular example, starting with a liquid solution of 80 per cent A, 20 per cent B, eventually gives the eutectic mixture L_E (comprising the 60 per cent A and 40 per cent B) surrounding dendrites of pure metal A, such that $A = L_E$. Thus starting with 1 lb of the liquid solution, gives 0.5 lb pure metal dendrites surrounded by 0.5 lb eutectic.   It is a simple matter to check that these figures are correct, since the eutectic contains $0.4 \times 0.5 = 0.2$ lb of metal B, as required.

Similarly, for the 30 per cent A, 70 per cent B liquid solution at temperature JKL, the proportions by weight of the pure metal B dendrites and liquid of composition J are:-

$$\frac{B}{L_J} = \frac{JK}{KL} \tag{2.8}$$

and on solidifying:-

$$\frac{B}{L_E} = \frac{EP}{PG} \tag{2.8a}$$

The curves CQE and DHE are known as *liquidus curves*, being the locus of temperatures above which the metal is liquid and the curve CFEGD (made up of three straight lines, in fact) is known as the *solidus*, being the locus of temperatures below which the metal is solid.

It is possible from considerations such as those just outlined to draw up a set of rules for use with equilibrium diagrams. In any two-phase area (such as CFEC in Fig. 2.13) the composition and relative amounts of the two phases present in an alloy of given composition at a given temperature under equilibrium conditions may be obtained as follows:-

(1)   Draw the vertical composition line (e.g. HKNP) and horizontal isothermal (e.g. MNO) concerned.
(2)   The intersection of the isothermal with the neighbouring phase boundary lines gives the composition of the two phases (e.g. *M* gives the composition of the liquid, *O* gives the composition of the solid).
(3)   The distances of the points of intersection with these boundary lines along the isothermal indicate the relative amounts of the two phases, e.g.:-

$$\frac{\text{Solid } B}{\text{Liquid of Composition } M} = \frac{\text{MN}}{\text{NO}}$$

This is often termed the *lever rule*, from the analogy of the isotherm MNO being a lever, with N as the 'fulcrum', and weights of the phases as 'weights'.

In the same way that the eutectic is an intimate mixture of two metals formed on solidification from a liquid solution, a *eutectoid* is an intimate mixture of the two metals formed from the breaking up of a solid solution. The eutectic or eutectoid structure is of great importance in determining the strength of the final alloy, in that the tensile strength and toughness are largely dependent on the mechanical properties of the metal which forms the continuous phase of the eutectic (the matrix in which the other metal is embedded). If this metal is brittle, the alloy will be brittle. The alloy of eutectic composition is that having the lowest melting point of the system and can therefore be very useful for soldering or brazing.

Typical alloys which are insoluble in one another in the solid state and therefore form eutectics are: lead-antimony, lead-tin, copper-silver, aluminium-silicon. An interesting alloy of non-metals which can be used to give a eutectic is that of ice and salt (NaCl), the liquid solution being brine. This solution has a eutectic composition of 23.3 per cent NaCl, 76.7

per cent H_2O, which solidifies to form an intimate mixture of ice and salt at a temperature of -6°F (-21°C). (It is not advisable to use this as anti-freeze in a motor vehicle, because it is rather corrosive, but it _is_ used on roads just because of its capacity to lower the freezing temperature of ice, so vehicles get corroded anyway!)

Binary alloys formed from metals which are completely soluble in each other in both liquid and solid states give equilibrium diagrams of quite different appearance. The metals copper and nickel are of this type, both having high melting points and similar (f.c.c.) lattice structure. Alloys of these two metals are formed by solidification into crystals of primary solid solution having only a slightly distorted lattice structure. (The atoms are of nearly the same radius as can be seen from Table 2.3: Cu = 1.275 Å, Ni = 1.25 Å.) Microscopic examination will not reveal more than the one type of crystal and it is impossible to tell one metal from the other in the alloy, unless there is more than 85 per cent copper present when the characteristic pink colour of copper becomes evident. Alloys of copper and nickel are customarily referred to as 'cupro-nickel' (up to 30 per cent Ni), 'constantan' or 'Eureka' (45 per cent Ni), and 'monel metal' (67 per cent Ni, 30 per cent Cu, 1.5 per cent Fe). The equilibrium diagram is sketched in Fig. 2.14.

Consider the cooling of a solution of 50 per cent Cu, 50 per cent Ni. Solidification begins at A, the isothermal cutting the liquidus at A and the solidus at B. Thus the liquid is of composition A, whilst the solid which first forms is of composition B, rich in nickel. As cooling proceeds down to the temperature represented by the isothermal CDE, the liquid becomes slightly depleted of nickel, at composition C, whilst the solid is still rich in nickel, and the relative weights of liquid and solid are given by the lever rule in the proportions:-

$$\frac{\text{Liquid of Composition } C}{\text{Solid of Composition } E} = \frac{\text{DE}}{\text{CD}} \qquad (2.9)$$

As the liquid slowly cools, copper atoms diffuse themselves into the solid so that the composition of the resulting solid travels down the line BEG, finally producing a solid solution of the same composition as originally possessed by the liquid solution. It is important to note that for any composition of the original liquid solution, the first alloy to solidify will always be richer in the higher melting point constituent. In practice equilibrium is rarely attained, due to the usually rapid cooling which occurs, and the crystals have a 'cored' structure, i.e. they possess a concentration gradient.

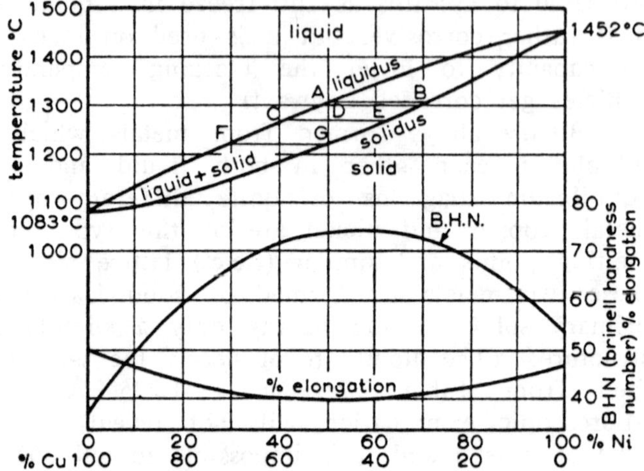

Fig. 2.14 Equilibrium diagram for mutually soluble metals
(copper and nickel)

This extends from the centre of the original dendritic skeleton to the outside, the core being richer in the higher melting point constituent (in this case nickel). (The cored structure may be removed by prolonged annealing [soaking] at a high temperature.) In this type of alloy forming a solid solution, the distortion of the crystal lattice due to the introduction of foreign atoms leads to increased resistance to deformation, i.e. increased hardness, which shows a maximum near the 50/50 alloy composition. This distortion may often be only very slight although leading to much greater hardness than for either pure metal.

The fact that the first liquid to solidify is richer in one of the constituents is made use of in the process of *zone-refining*. This process is used for purifying metals, particularly germanium and silicon for use in transistors, and consists in causing a melted zone to travel along a bar of the impure metal so that the solid frozen out becomes purer than the average composition, whilst impurities are deposited at the end of the bar.

In practice the simple basic types of phase diagram just described rarely occur singly; one of the simplest examples of a combined eutectic and solid solution diagram is that for the lead–tin solder alloy shown in Fig. 2.15. The liquidus curves are AE and EC, the solidus ABEDC. Thus the areas shown shaded between AE and AB, and between EC and DC, resemble areas between the liquidus and solidus of Fig. 2.14.

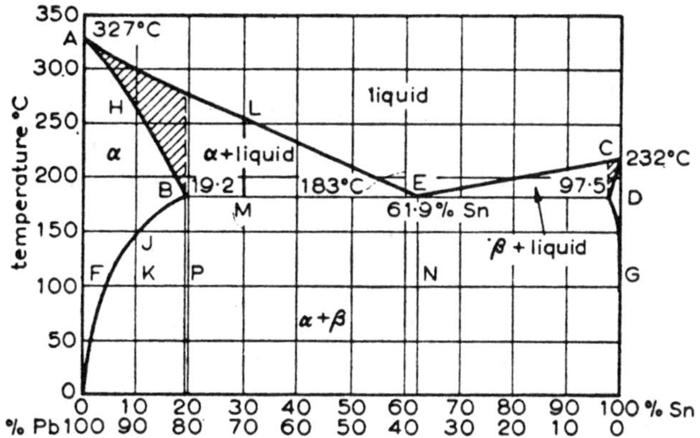

Fig. 2.15 The lead-tin system

Alloys containing less than 19.2 per cent of the tin will form a solid solution (called α) and less than 2.5 per cent of lead will form a solid solution called β. Considering the 10 per cent tin alloy, for example, it will form a solid solution on cooling below the temperature H (which will be cored unless cooling is very slow). Such behaviour may be termed partial solid solubility, and the customary convention of denoting such phases by Greek letters, e.g. α-phase, β-phase, γ-phase etc., should be noted.

The lines FB and GD on the diagram represent phase changes which occur in the solid state. In this case they represent the limits of solid solubility of the one metal in the other. Thus the point J indicates the temperature below which 10 per cent tin cannot be accommodated in solid solution in the lead. Below this temperature, for example at 100 °C, the primary dendrites of composition J will split up into two solid solutions of composition F and G in the proportion by weight of the lengths KG to KF. Similarly, considering the 30 per cent tin alloy, solidification will commence at the temperature L and when cooling has continued until the eutectic temperature has been reached at point M, the melt will consist of α dendrites of composition B surrounded by liquid of composition E (61.9 per cent tin) in the proportions of the lengths ME to MB. On further cooling the liquid will solidify to form a eutectic mixture of the two solid solutions of α (composition B) and β (composition D).

Further cooling, down to 100°C, say, will result in the gradual splitting up of the primary α dendrites of composition B into solid solutions of composition F and G, in the proportions PG to PF where P is vertically below B. The solid solutions forming the eutectic mixture will also gradually change their composition from α_B and β_D to α_F and β_G (where α_B means a solid solution of composition given by the point B, etc), and their proportion by weight will change from α_B/β_D = ED/EB to α_F/β_G = NG/NF, where N is vertically below E. All these changes in the solid state take considerable time to complete, and may be completely prevented by rapid quenching.

Many pairs of metals form *chemical* or *valency compounds* which obey the usual valency laws, or *electron compounds* which do not. Such compounds are given the generic term *intermediate phases;* an example of the first type is Mg_2Sn, and the second Cu_5Si. Equilibrium diagrams for a binary alloy of the first type consist simply of two eutectic diagrams placed together, but the second type of intermediate phase leads to complicated phase diagrams due to the variable composition of the electron compound possible in any alloy.

A number of dissociation reactions, all of which are reversible, may take place due to the relative instability of complicated crystal structures at high temperatures. As the temperature is raised the increased thermal vibrations tend to dissociate the structure into simpler forms such as:-

(1) Peritectic: solid A → solid B + liquid
(2) Peritectoid: solid C → solid D + solid E
(3) Calcination: solid F → solid G + gas.

The first of these is of the most interest in relation to metals, an important example being found in the brass alloys (copper-zinc). The copper end of the equilibrium diagram is illustrated in Fig. 2.16. The α constituent possesses a face-centred cubic lattice, as for copper, the zinc atoms replacing copper atoms in the lattice. The β constituent has a body-centred cubic structure and will only exist at room temperature when the composition lies between narrow limits (46-50 per cent zinc). Any alloy containing up to 32.5 per cent of zinc or more than 38.5 per cent zinc will solidify in the normal way, although the resulting solid solutions are different (either α or β respectively). An alloy containing 35 per cent Zn will start to solidify at temperature L and will cool down to M, at which point there will be α dendrites of composition D surrounded by liquid of composition B, in proportion MB to MD by weight.

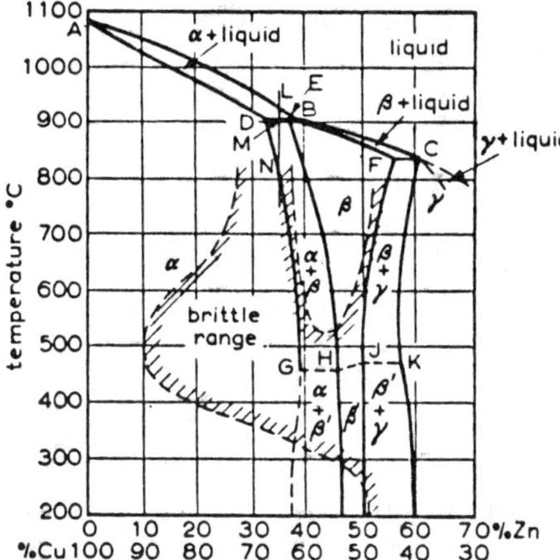

Fig. 2.16 The copper-zinc system (brass)

The peritectic reaction then occurs, in which the liquid B reacts with the solid α_D, producing a new solid β_E, so that we are finally left with an alloy mixture of solid α_D and β_E crystals. The peritectic reaction may be regarded as the inverse of a eutectic reaction, since it involves a change from two phases to one phase, whereas the eutectic reaction involves a change from one phase to two phases on cooling. During cooling to N the relative proportions of the α and β crystals change, the β crystals gradually disappearing so that the α crystals become richer in zinc until only solid crystals of α_N remain. Alloys containing between 38.5 and 46 per cent of zinc solidify as β crystals but on crossing the curve EH some α crystals will be formed, separating out first at grain boundaries then in needle plates along crystallographic planes within the β crystals. This type of structure is often referred to as a *Widmanstätten* structure and occurs in the well-known 60/40 brass (Muntz metal). Below the line GHJK a rearrangement of the atoms takes place, to give a *superlattice* type of structure. When the Zn content exceeds about 50 per cent, a third constituent (γ) forms a continuous film around the β crystals. It is hard and very brittle, leading to the possibility of easy fracture so that such alloys are rarely used in practice.

An important industrial metal, not only because of its cheapness and fluidity in the molten state but also because of its high compressive strength under static loading, is *cast iron*. Cast iron is essentially remelted pig iron of about 95 per cent purity containing from $2\frac{1}{2}$ to 4 per cent carbon, and up to 3 per cent silicon, together with small amounts of manganese, sulphur, and phosphorus. The iron-carbon equilibrium diagram is sketched in Fig. 2.19 and is an important example of the existence of both eutectic and eutectoid points. As discussed in the next section, the various lattice structures formed by iron and carbon have characteristic names, austenite being a f.c.c. lattice, ferrite a b.c.c. lattice and cementite an orthorhombic lattice structure which may be represented by the symbols Fe_3C. The eutect*oid* mixture of ferrite and cementite is called *pearlite*, whilst the eutect*ic* mixture of austenite and cementite is called *ledeburite*.

Considering an alloy containing 3 per cent carbon on solidification, dendrites of austenite are deposited out until the eutectic temperature of 1130°C is reached. On cooling through this temperature the remaining liquid solidifies into ledeburite, and on further cooling down to the eutectoid temperature of 725°C both the austenite dendrites and the austenite in the eutectic mixture give up carbon to form cementite. At 725°C all the austenite, which is now of eutectoid composition (0.87 per cent carbon), transforms into pearlite, so that the final metal consists of an intimate mixture of ferrite and cementite, both of which have a white appearance and therefore give a white fracture. This is a *white cast iron*, a high proportion of the alloy being cementite which is hard and brittle, so that the alloy itself is hard and brittle and difficult to machine.

Cementite is not as stable in iron as is graphite, and it will dissociate into iron and graphite if given sufficient time. Although this dissociation process is usually undetectable, it may be accelerated by prolonged heating in the presence of more than approximately 1 per cent silicon as a catalyst. The majority of cast irons have more than $1\frac{1}{2}$ per cent silicon present from the pig iron so that graphite forms during the solidification process. The iron-graphite equilibrium diagram, although very similar to that for iron-cementite, is slightly different (depending on the amount of silicon present) as indicated by the broken lines in Fig. 2.19, and stable graphite appears in every area in place of cementite. It is possible, in fact, to define a per cent *carbon equivalent value* = per cent carbon + 1/3 (per cent silicon + per cent phosphorous). In this way the effect of these elements may be allowed for without having to modify the phase diagram. Thus the cast iron resulting on cooling from the liquid in the presence of silicon will contain ferrite and graphite, both of which are soft, the graphite being present in the form of coarse flakes, which are grey in colour. This is a *grey cast iron*, and is much softer than white cast iron, but still lacks ductility and shock-resistance due to the presence of the graphite in the form of coarse flakes which act almost as voids within the structure of the material, thereby producing stress concentrations under load.

The addition of small quantities of magnesium or cerium to the melt immediately before casting causes the graphite to form in spheroidal nodular particles. The resulting *nodular cast iron* has markedly improved mechanical properties, the percentage elongation being increased from about 1 per cent up to about 15 per cent. A type of nodular graphite structure may also be produced by heat-treating a white cast iron, formed from an alloy of low silicon content by rapid cooling. Prolonged annealing of this white cast iron causes dissociation of the cementite which is deposited as *temper carbon* in the form of clusters of graphite resembling the spheroidal graphite obtained by the addition of magnesium or cerium, and similar ductility is obtained. The resulting product is called *malleable cast iron* and is made either from the *Black Heart* process (U.S.A.) or *White Heart* process (Europe).

Before leaving this general discussion of phase diagrams, mention should be made of the method of representing the equilibrium behaviour of ternary alloys using phase diagrams. As might be expected, a ternary equilibrium diagram must be made up from a number of *surfaces*, rather than *lines* as for binary alloys. It is nevertheless usual to plot the diagrams on the basis of an equilateral triangle, whose sides represent the metals concerned, as illustrated in Fig. 2.17.

Now the sum of the perpendicular distances from any point P within the equilateral triangle to the three sides is constant and equal to the height of the triangle. Thus for the point P shown in Fig. 2.17 this fact may be used to designate the composition 35 per cent *A*, 21 per cent *B*, 44 per cent *C*. The addition of a third metal to a binary system often results in still further lowering of the temperatures at which eutectics occur. It is difficult to represent a three-dimensional surface in a plane diagram, and what is often done is to draw the isotherms as contours of a 'map' of the particular surface (e.g. liquidus or solidus) of interest. It is clear from Fig. 2.17 that the vertical faces above each side of the basic equilateral triangle will constitute and contain the equilibrium diagram for the appropriate binary alloy, e.g. if *a* is zero the point P will move along the side of the triangle between 100 per cent B and 100 per cent C.

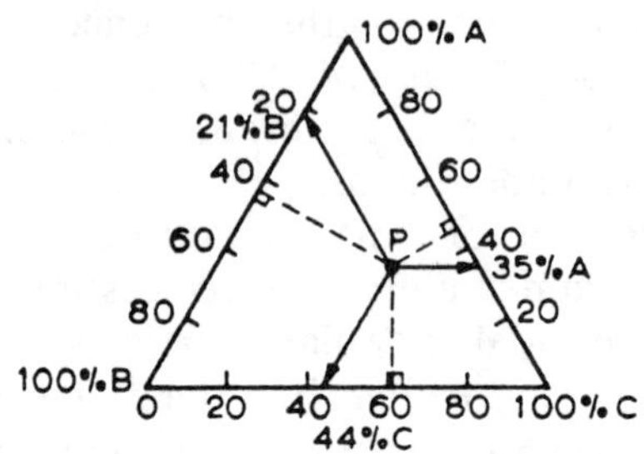

Fig. 2.17 Section of a ternary diagram

Similarly it is possible to take vertical sections through a complete ternary diagram to give binary phase diagrams for fixed percentages of the third constituent (by taking sections parallel to one of the sides of the diagram). The intersections of liquidus surfaces often result in the representation of the eutectic by a line in a ternary system. Such eutectic lines may meet at a *eutectic point* for which the ternary alloy composition gives the minimum eutectic temperature. Although slightly more complicated, a similar 'lever' principle can be used for determining the relative proportions of the constituents in a ternary alloy from the equilibrium diagram.

2.3 The Effects of Heat Treatment

2.3.1 *Solid Solutions and Precipitation*

Behaviour of the type illustrated by the lead-tin alloy with 10 per cent tin (the vertical line HJK in Fig. 2.15) is very important in heat treatment processes. *Solution treatment* involves heating along this line so that the two phase alloy made up of the two solids α and β is converted into the single phase α alloy; the solid β is in fact *dissolved* in the solid α as the metal is heated. The process is comparatively slow since time is required for the necessary solid diffusion to occur. The converse cooling cycle occurs even more slowly since the temperatures are lower, and is referred to as *solid precipitation*, since the β phase is being precipitated out of the solution. Quite often a noticeable increase of hardness occurs during the initial stages of precipitation (if the precipitated phase has suitable properties).   In practice precipitation may take considerable time to occur at room temperature, so that the resultant hardening is often referred to as *age hardening*. If this treatment is applied at elevated temperature it is called *precipitation hardening*.

2.3.2 *Age hardening*

In commercial heat treatment, which utilizes this phenomenon to harden metals, the alloy having been solution treated is quenched through the phase change (line FJB of Fig. 2.15, for example), to produce a *supersaturated solid solution* (supersaturated since the equilibrium solubility decreases with decreasing temperature).   The supersaturated solid solution is initially quite ductile, but age hardens as precipitation of the β phase occurs.   Thus the component (aluminium air-frame rivets are an example) may be worked easily into the required shape initially, becoming harder and more resistant to deformation as time progresses.   The age-hardening process is accelerated by heating, such treatment being referred to as *artificial ageing*.   If it is required to store the components before working, ageing may be inhibited by refrigeration, so that close control of material properties is possible.   One of the most important alloys exhibiting this behaviour is the aluminium alloy containing about 4 per. cent copper (the old trade name 'Duralumin').   The aluminium end of the aluminium-copper phase diagram is sketched in Fig. 2.18.

The solution treatment temperature usually used is approximately 490°C, after which it is generally aged at room temperature for about four days; higher strength can be obtained with artificial ageing, although experience shows that this is at the risk of making the alloy subject to inter-crystalline corrosion.   The θ constituent referred to in Fig. 2.18 is an intermediate phase based on the chemical compound $CuAl_2$, and the eutectic is an intimate mixture of α and θ. (No eutectic occurs in alloys of less than 5.6 per cent copper, of course.)   At room temperature the structure of the 4 per cent Cu alloy consists of crystals of α aluminium containing 0.5 per cent of copper in solid solution, the remainder of the copper having been precipitated out, eventually to form finely dispersed particles of $CuAl_2$.

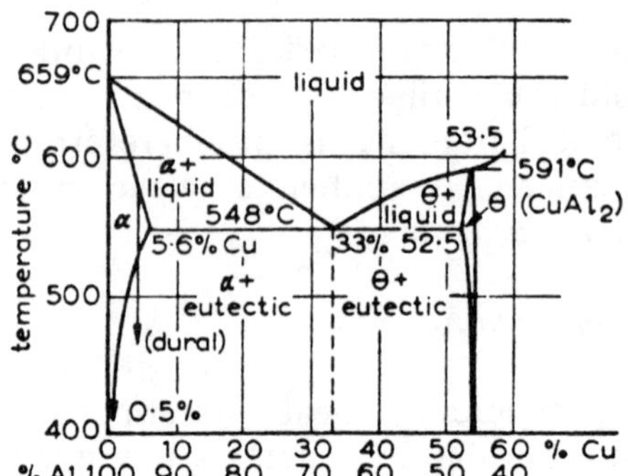

Fig. 2.18 The aluminium-copper system

This alloy is eminently suitable for solution treatment, quenching, precipitation and age hardening, therefore, with certain precautions. The solution treatment temperature is kept below the eutectic temperature to eliminate any possibility of fusion or melting. Since increase of either time or temperature of ageing accelerates nucleation and coalescence of the fine precipitate, *over-ageing* is possible and the treatment must be carefully controlled to avoid this. As discussed later, the mechanism of hardening is by obstruction of the movement of dislocations, caused in this case by the finely dispersed precipitate, and coalescence of these particles reduces the number of obstructions present.

In the general case of age hardening, most alloy systems exhibit the phenomenon of firstly forming zones or clusters of atoms of the element being precipitated (copper atoms for Duralumin), which are still part of the parent lattice. On continued ageing there may be one or more intermediate precipitates formed in these zones before the equilibrium precipitate ($CuAl_2$ for Duralumin) is finally formed, exhibiting less and less coherence with the parent lattice structure. The zones are referred to as Guiner-Preston or G.P. zones, and are initially one or two atomic layers thick in the aluminium-copper alloy. Maximum hardness is obtained with one of the intermediate precipitates (θ' or θ'' in Duralumin), being due to the straining or distortion of the lattice. Over-ageing converts the intermediate precipitate into the equilibrium precipitate (θ or $CuAl_2$ for Duralumin) which completely lacks coherence with the matrix lattice and forms less of an obstruction to the movement of dislocations. This lack of coherence of the precipitate is an extremely important phenomenon, leading as it does to a considerable reduction in the strength and toughness of the resulting material.

The shape of the initial zones also affects the amount of hardening; spheres are most effective, discs or platelets least effective, with needles intermediate. In some alloys, *discontinuous precipitation* may occur at grain boundaries which may or may not be beneficial depending on the application for which the alloy is required. *Dispersion hardening* occurs in some systems, notably the aluminium-zinc-magnesium alloys, in which very fine precipitates are formed with apparently no resultant straining of the parent lattice and yet giving increased hardness.

2.3.3 *Quenching and Tempering Steels*

The principles just discussed for precipitation hardening are very similar to those which apply to the tempering of quenched steels, which will now be considered. As is well known, steels and cast irons are both 'alloys' of iron and carbon; carbon, although not a metallic substance, possesses a small atomic radius (approx. 0.71) and can therefore be accommodated within the interstices of the iron crystal lattice in the face-centred cubic form (but only with difficulty in the body-centred cubic arrangement). The iron end of the iron-carbon diagram is shown in Fig. 2.19.

Pure iron changes its lattice structure from body-centred cubic to face-centred cubic on heating above 906°C, changing back to body-centred cubic at 1403°C. This type of behaviour, namely the existence of a pure substance in more than one crystalline form, each of which is stable over certain limits of temperature (and ·pressure), is termed *allotropy*, or *polymorphism*, and is exhibited by many metals. (Uranium has three and plutonium six allotropic forms, each of different crystal structure, and therefore fairly large volume changes are involved on change of temperature.) As can be seen from the iron-carbon equilibrium diagram, the solubility of carbon in the b.c.c. modification of pure iron (0° - 906°C and 1403° - 1537°C) is very low, but in the f.c.c. form (*austenite* region) it is somewhat larger. The b.c.c. structural form of pure iron which exists at room temperature is called α-iron or *ferrite*, is soft and ductile and ferromagnetic below 770°C (the Curie point). (The non-magnetic α-iron was originally known as β-iron until it was discovered that there was no change of crystal structure.) The f.c.c. form of pure iron is called *γ-iron* or *austenite*, which is only stable over the high temperature range indicated on the diagram, but is soft, ductile, and easily worked at those temperatures. Austenite is not ferromagnetic at any temperature.

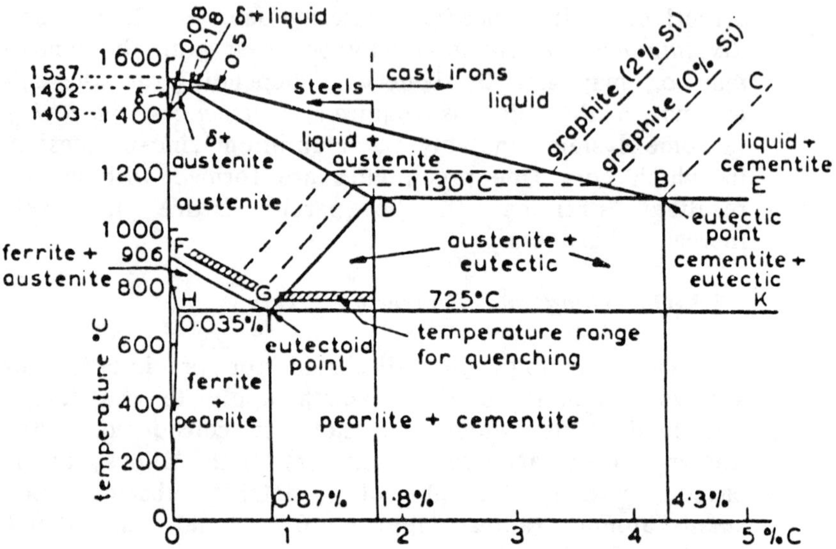

Fig. 2.19 Part of the iron-carbon diagram

Above 1403°C the austenitic form of pure iron cannot exist,
and the crystal lattice reverts to the b.c.c. form. This
modification of pure iron is called δ-*iron* or δ-*ferrite,* since it
is the same as ferrite except for its stable temperature range.
The solid solutions formed by carbon in α, δ, and γ iron are
also referred to as ferrite, δ-ferrite, and austenite, so that
zones are formed as shown in the diagram.

By customary definition, *steels* are iron-carbon alloys
possessing less than 1.8 per cent carbon, whilst alloys
possessing more than 1.8 per cent, and less than 6.7 per cent
carbon, are *cast irons.* Iron and carbon form a eutectic
(ledeburite) at point *B,* the eutectic mixture being austenite
and cementite. *Cementite* has the chemical composition of
Fe_3C and may be called *iron carbide,* although this is not
desirable as the elements do not in fact form molecules as for
a normal chemical compound, but exist in a crystal lattice
containing iron and carbon atoms in a 3:1 ratio. The unit
cell is orthorhombic with 12 iron and 4 carbon atoms, giving
a carbon content of 6.68 per cent (by weight). Compared
with ferrite and austenite, cementite is very hard, and will in
the spheroidal form greatly increase the strength of steel when
dispersed in a soft matrix of ferrite. The eutectic is of
importance in the formulation of cast irons as already

discussed.

The point G on the diagram is a eutectoid point, of importance in the formulation of steels, and indicates the composition of (solid) austenite which on slow cooling will change into a mechanical eutectoid mixture of ferrite and cementite, known as *pearlite,* due to its resemblance to mother-of-pearl in a polished and etched specimen. The structure of pearlite is lamellar, being formed from alternate plates of ferrite and cementite, there being about 87 per cent ferrite and 13 per cent cementite present, and the ferrite forms a continuous soft matrix in which the cementite plates are dispersed. As remarked previously, this dispersion of 'disc'-like precipitate does not give much toughness to the material. The eutectoid reaction commences at the boundaries of the austenite grains where nucleation of the new phases occurs most readily. Carbon atoms diffuse from the f.c.c. austenite areas, which will become the ferrite matrix, into the lamellar zones, which will become cementite, the f.c.c. structure changing into b.c.c. iron and orthorhombic cementite, respectively. This diffusion process requires time to occur and the pearlite 'grows' inwards with the layers edgewise from the grain boundaries, which may still be seen under the microscope after the transformation has been completed.

A steel containing 0.87 per cent carbon is known as a *eutectoid steel;* less than 0.87 per cent *hypo-eutectoid,* beyond 0.87 up to 1.8 per cent carbon *hyper-eutectoid.* It may help to note that the word hyper- is from a Greek word meaning 'beyond'. Hypo-eutectoid steels formed by slow cooling are tougher than hyper-eutectoid, since they have ferrite separating areas of pearlite rather than the brittle cementite which is deposited along the grain boundaries.

Considering the effect of rate of cooling on steels, the situation is analogous to that of precipitation hardening just described, although somewhat more complicated because of the existence of the eutectoid reaction. The steel is first made austenitic by 'solution' treatment at a suitable temperature, as indicated in the figure. (Since the cementite present in hyper-eutectoid steel below the austenite region can contribute to the final hardness, such steels are not fully austenitized, significantly reducing the danger of cracking on quenching.) To simplify the discussion, consider a eutectoid steel. The *rate* of cooling may be altered by the typical treatments shown in Table 2.4.

TABLE 2.4

Rate of cooling	Final structure	Appearance
Slowest— Furnace-cooled (annealing)	Pearlite	Laminated
Air-cooled (normalizing)	Pearlite	Finely laminated
Air-blast	Sorbitic Pearlite	Very finely laminated
Oil-cooled	Sorbite	Extremely fine lamellae (hardly visible)
Hot water	Troostite	Nodular, each nodule formed from radiating lamellae (hardly visible)
Warm water	Troostite and Martensite	
Fastest - Cold water	Martensite	Needle-like or acicular Structure

The effect of increasing the cooling rate is to produce progressively harder material, each product bearing a characteristic name, as shown in Table 2.4. Martensite is the hardest form of steel which is obtained by very rapid quenching, and is a supersaturated solid solution of carbon in α-iron, existing in the form of needles, the crystal structure being body-centred tetragonal. This is a b.c.c. structure having one edge longer than the other two and arises from the fact that the carbon atoms are still present in what would otherwise be the b.c.c. ferrite lattice. The resultant distortion of the lattice makes it resistant to the movement of dislocations and therefore very hard and brittle. It is not possible completely to retain the *austenitic* structure in straight carbon steels, no matter how fast the quenching rate, but it is possible to retain it by adding alloying elements. (In particular, elements such as nickel and manganese are more soluble in the f.c.c. γ-form of iron, and therefore increase the temperature range over which the austenite is stable; chromium and tungsten are more soluble in the b.c.c. form of iron and decrease the stable γ temperature range.) *Bainite* is another modification which is more often found in alloy steels, and has an acicular structure.

The effect of temperature and cooling rate in the heat-treatment of steels has been studied extensively, and much information exists in the form of isothermal transformation diagrams (alternatively known as Time Temperature Transformation diagrams, TTT or triple T diagrams, or S-curves, because of their characteristic shape). These diagrams are constructed from the results of experiments in which small specimens of steel are austenitized and then cooled virtually instantaneously in a suitable liquid bath at the temperature under investigation and the transformation is allowed to proceed at that temperature. By continually measuring the hardness or dilatation of the specimen at various stages in the transformation it is possible to determine the initiation and completion of the transformation (which has occurred isothermally) and a curve may be plotted of transformation temperature versus transformation time. Such curves display at the nose or knee of their S-shaped curve a minimum transformation time at a temperature only 200 C° or so below the equilibrium phase change temperature, fine pearlite being formed. Below this temperature, down to about 250 °C, the times for isothermal nucleation and completion of the transformation increase considerably and bainite is formed. Below this temperature martensite forms almost immediately on quenching, although the amount formed depends on the temperature.

In practice, treatments are rarely isothermal but, although not strictly permissible, an idea of the behaviour can be obtained by superimposing the cooling curve on the TTT curve for the particular steel concerned. Examples are shown in Fig. 2.20, which is taken from the book by ANDERSON et al. A phenomenon analogous to precipitation hardening is produced in steels by the process of tempering after quenching. Providing the quenching has given a sufficiently rapid cooling rate (and this may be difficult for heavy steel sections), the supersaturated solid solution of carbon in iron (martensite) will have been formed. If the temperature is now raised (to say 100 °C) the martensite will break down to form an intermediate carbide which is initially coherent with the matrix. If the temperature is raised still further the remaining low carbon martensite (and another constituent, ϵ-carbide) will be converted into ferrite and non-coherent cementite, present eventually as globular particles. The higher the temperature the more rapid will be the precipitation of globular cementite, and there will be a greater tendency for the precipitated particles to coalesce, leading to reduced hardness and increased ductility.

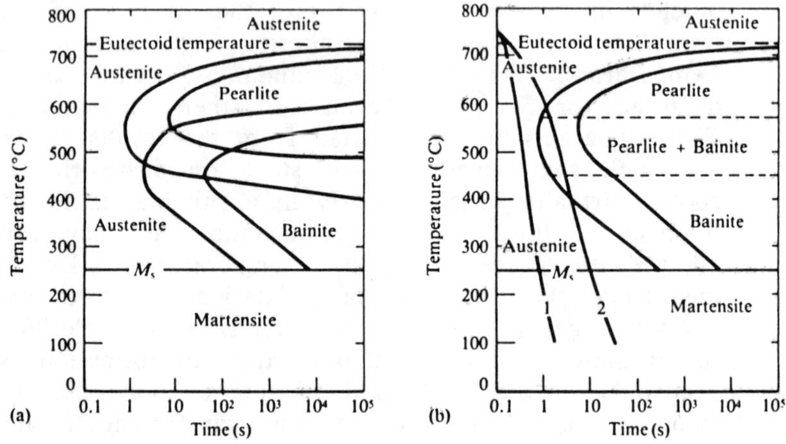

Fig. 2.20
(a) Schematic diagram of the overlapping TTT curves for the pearlitic and bainitic transformations in the plain carbon steel of the eutectoid composition (0.8 wt.%C).
(b) TTT diagrams for the same steel showing a single, continuous curve with two superimposed cooling rates designated 1 and 2

Steels are therefore always used in the 'over-aged' condition, and the best combination of properties for any particular application can only be obtained by careful control of the tempering temperature. As stated previously, martensite is the hardest form of steel, and as the tempering temperature is increased the hardness falls, but the ductility and impact strength increase. Thus tool steels are customarily tempered at about 300°C, whilst structural steels are tempered at about 600°C to give higher ductility and resistance to shock at the expense of some hardness (or strength).

In practice plain carbon steels suffer from certain limitations, for example, they require cold water quenching to produce martensite, which tends to introduce quenching stresses which may cause distortion and/or cracking. Thus the section thickness which can be hardened efficiently is limited to about ½in. In addition, such steels are deficient in special properties such as resistance to creep and corrosion. For these reasons alloy steels have been developed, since by the addition of alloying elements it is easier to form martensite, bainite, or austenite.

The production of bainite has been developed into a commercial process called *austempering*, in which the cooling rate is controlled so as to transform the austenite isothermally into bainite, which is not so brittle as martensite and does not require such dangerously high cooling rates for its formation. Alternatively the danger of cracking may be alleviated by using an interrupted quench, or *marquenching*, in which the steel is quenched to just above the critical temperature for martensite formation (about 250°C) and then slowly cooled to form martensite. The initial quench must be rapid enough to avoid the pearlite transformation. The lower bainite transformation is slower so that there is time for the temperature to equalize before final cooling to room temperature. These two treatments are particularly useful with large masses for which a water quench might cause cracking but slower quenching would not produce sufficient martensite, and for which a highly alloyed steel is not suitable, perhaps because of expense.

On tempering martensite, the resulting cementite or iron carbide may be transformed by diffusion of the alloying elements into it (a process known as *secondary hardening*) to form alloy carbides which are often more stable than the cementite, especially at high temperatures, and also serve as hardening precipitates to give good creep resistance. Such alloying elements considerably reduce the rate of cooling necessary to produce martensitic structures, so that sections up to 4 in. thick can be hardened. Alloy steels suffer from certain disadvantages, however; they are more expensive to produce and suffer from the phenomenon of *temper brittleness*, in that the tempered alloy steel has a low impact strength possibly due to the segregation of solute atoms on grain boundaries. It is found that the addition of a small amount (0.75-0.5 per cent) of molybdenum reduces this tendency.

ROLLASON[1] suggests a useful grouping of alloying additions to steel as follows:-

(1) *Elements which tend to form carbides:*
Chromium, tungsten, titanium, columbium, vanadium, molybdenum, and manganese. (The mixture of complex carbides is often referred to as cementite.)
(2) *Elements which tend to graphitize the carbide:* Silicon,

cobalt, aluminium, and nickel. (Only a small quantity of these elements may be added without deleterious effect.)
(3) *Elements which tend to stabilize austenite:* Manganese, nickel, cobalt, copper, and zirconium (all f.c.c.).
(4) *Elements which tend to stabilize ferrite:* Chromium, tungsten, molybdenum, vanadium, and silicon (all b.c.c.).

These last elements diminish the amount of carbon soluble in the austenite, and may even cause the austenite phase to disappear completely. Thus 18 per cent chromium irons cannot be heat-treated by quenching and tempering at all. The effects of each of these four groups of alloying additions can mutually interact, so that it is possible to devise alloy steels suitable for any required special application.

As pointed out by ROLLASON[1], an important feature of the precipitation hardening reactions in both ferrous and non-ferrous alloys is that the optimum engineering properties are generally obtained before the precipitate is visible under the optical microscope.

2.3.4 *Annealing and Normalizing*

Annealing is a generic term which may have several meanings. In alloys that do not undergo a phase change on heating it is most often used to denote the process of *recrystallization* of a metal which has been cold-worked so that the atomic lattice is in a strained condition. The strained lattice possesses more energy than it would in the unstrained configuration, to which it will readily revert on heating or annealing. Hardness is a good indication of the occurrence of recrystallization and the temperature at which marked softening occurs is called the recrystallization temperature. It is found that a more heavily worked metal will soften at a lower temperature, since the atoms already possess more energy. As shown by VAN VLACK[5], the *homologous* (or *"corresponding"*) temperature for recrystallization T_H (= T/T_M, where T_M = melting temperature) is approximately 1/2 for most metals. If the temperature is raised these new crystals will coalesce, a phenomenon known as *grain growth*. Recrystallization is also markedly time dependent.

When used in connection with the heat treatment of steels that undergo the eutectoid reaction, annealing involves heating the steel into the austenitic phase and allowing it to cool very slowly in the furnace. If the steel is allowed to cool in air the process is referred to as normalizing, and the microstructure of the resulting pearlite is considerably finer than in an annealed steel, due to the more rapid cooling. *Process annealing* is often used to relieve stresses in cold worked steels without forming austenite, and involves heating to just below the eutectoid temperature for a short time. Stress relief of this type is often called *recovery*, although it may be carried out to recrystallize, in low carbon steels.

An annealing treatment often used to achieve better ductility is that of *spheroidizing*. This is really an extended subcritical anneal carried on for a sufficient length of time to allow the coalescence of any carbides present into spherical particles. Considering the lamellar form of cementite present in pearlite, a long period of heating to convert it into *spheroids* is required, whereas only a relatively short period is required for coalescence of the already spheroidized tempered martensite. It is therefore better carried out on a quenched and tempered rather than annealed structure. In each case the spheroidite type of structure is more ductile than either the lamellar pearlite or the acicular martensite.

2.4 The Role of Dislocations and Other Defects in the Lattice

2.4.1 *The Mechanism of Plastic Deformation*

The difference between elastic and plastic deformation on the macroscopic scale, as evidenced by the stress-strain behaviour observed in the standard tensile test, has been discussed in Chapter 1. To understand more clearly the behaviour of metals subjected to large deformations it is helpful to study, at least qualitatively, behaviour on the atomic scale. It helps considerably in the initial discussion of this problem to restrict attention to a single crystal.

During elastic straining of a single crystal, which is made up of a regular array of atoms as already described, the interatomic forces readjust themselves to balance the applied stress. This is achieved by a slight displacement of the atoms in the lattice, which, of course, leads to the small measured distortion observed in practice.

For plastic deformation to take place it is generally necessary for large-scale *slipping* of adjacent planes of atoms to occur. The structure of the crystal lattice is maintained, but atoms change their relative places. Each different kind of lattice structure (e.g. b.c.c., f.c.c., h.c.p.) has planes along which such slipping can occur, and definite possible directions in which the relative slip may take place.   These are known as *slip planes* and *slip directions* respectively.   Slip occurs on relatively few planes, and the movement of the metal appears to occur in 'blocks' which form a series of steps on a polished surface showing up as black *slip bands* under the light microscope.   Slip is sometimes referred to as *glide*, and slip bands as glide lamellae; it takes place along the available direction in which the resolved shear stress is greatest.   This behaviour is illustrated for a circular tensile specimen of a single crystal in Fig. 2.21, each slip band being made up of several glide lamellae.   Slipping does not take place over the whole slip plane simultaneously — each atom moves individually, the movement being propagated across the plane in a manner described in the next section.   In both the hexagonal close packed and face centred cubic structures the atoms are arranged in their most closely packed configuration. The planes of closest packing are the planes along which slip is easiest, since the distance between such adjacent planes is greatest.   For a similar reason the easiest slip directions are those in which the atoms are most closely packed.

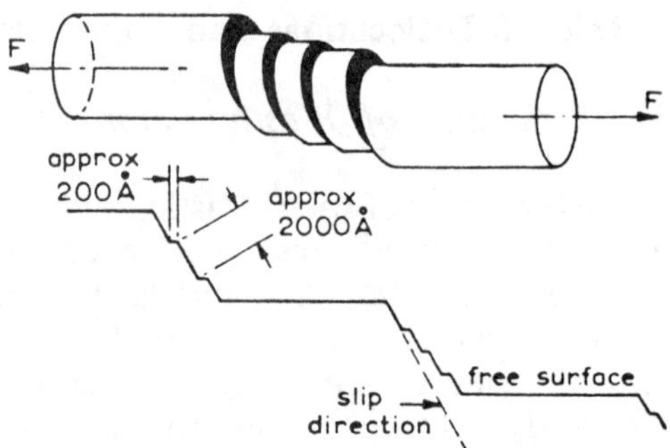

Fig. 2.21 Slip mechanism in tension of a single crystal

Because of the similarity of these two structures they are illustrated together in Fig. 2.22, the slip planes and directions being as indicated.   It may be observed that in a single crystal of the hexagonal array there is only one type of slip plane available, although there are three possible slip directions associated with it.   On the other hand there are four close-packed planes available in the f.c.c. crystals with three slip directions associated with each, giving a total of twelve possible slip systems.

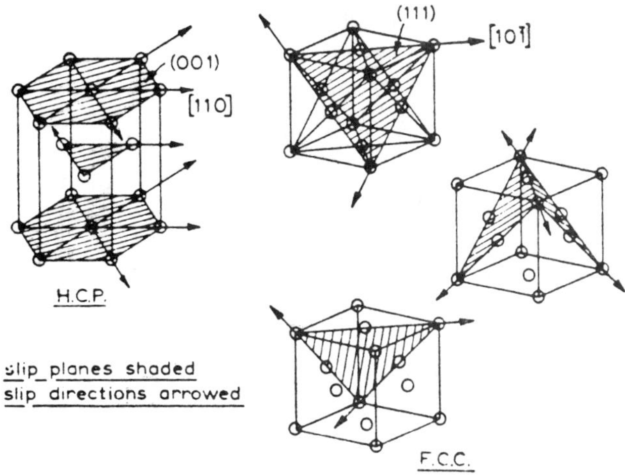

Fig. 2.22 Slip planes and slip directions
in h.c.p. and f.c.c. structures

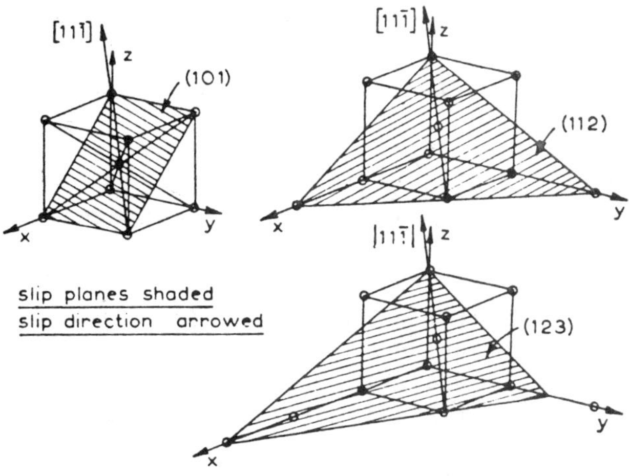

Fig. 2.23 Slip planes and slip directions
in b.c.c. structure

In the body centered cubic structure there are *no* close
packed planes, although the cube diagonal (11Ī) is a close-
packed direction, which is, therefore, always the slip direction.
Possible slip planes in this structure are indicated in Fig. 2.23,
and all of these can occur in the ferrite crystal, giving a total
of 42 possible slip systems.   In other b.c.c. metals the number

of planes available is generally less.

Obviously, the strength of a single crystal must be framed in terms of the relation between the shear stress and shear strain occurring on the slip plane being considered, and the *critical shear stress* at which gliding occurs on such a plane is found to be constant and independent of the crystal orientation, usually being very low {between 20 and 200 lbf/in^2, or 0.138 to 1.38 N/mm^2 (MPa)}. Considering a prismatic tensile specimen of a single crystal, if P is the applied load, A the cross sectional area, α the angle between specimen axis and slip plane, β the angle between specimen axis and slip direction, then:-

Area of slip plane = $A/\sin \alpha$
Resolved load in slip direction = $P \cos \beta$,
Therefore, the resolved shear stress in the slip direction is:-

$$\tau = \frac{P}{A} \sin \alpha \cos \beta \qquad (2.10)$$

Also it is possible to show that the shear strain γ may be determined from the expression

$$\left[\frac{l_1}{l_0}\right]^2 = 1+2\gamma \sin \alpha \cos \beta + \gamma^2 \sin^2 \alpha \qquad (2.11)$$

where l_0 and l_1 are the lengths of crystal before and after deformation (see PASCOE[6], p.131).

Metals can also deform by a process known as *twinning*. Each plane of atoms moves a definite distance in relation to a crystallographic plane which is characteristic of the particular lattice structure and separates the displaced from the undisplaced portion of the crystal. The movement is such as to cause the resulting atomic lattice to form a mirrored image at the *twinning plane* [which is (111) for f.c.c., (112) for b.c.c., and (102) for h.c.p.]. Thus the arrangement of atoms in the close-packed planes already referred to in the f.c.c. lattice - *ABC ABC ABC* - becomes on twinning - *ABC ABC BAC BAC*. The direction in which relative movement occurs is referred to as the *twinning direction* [112] for f.c.c., [111] for b.c.c. Twinning is an abrupt process which requires a higher stress for its initiation than does deformation by glide, and is accompanied by an audible noise, known as "tin cry". It rarely occurs due to deformation in the f.c.c. crystals, although twins can be formed in cold worked f.c.c. crystals by annealing (known as *annealing twins*), so that their presence indicates that the annealed metal had been previously cold worked. Thus twins due to deformation are often referred to as *strain twins*. Twins can be produced in ferritic iron by rapid deformation (e.g. a hammer blow) at room temperature, or by slower deformation at low temperature, such twins being known as *Neumann bands*. In spite of the abruptness of the occurrence of twinning, it occurs by step-wise atomic displacements in much the same way as glide, as explained by COTTRELL[3], p.87.

The present discussion has, of course, been over-simplified – in practice both slipping and twinning may occur in several crystals of a polycrystalline metal and in three dimensions. Severe deformation causes a breaking up of the original crystal structure, so that sub-grains are formed, the continuity of the lattice is broken, and the metal becomes more resistant to the glide (work hardened). Finally, the stress required to cause further glide becomes so high that it attains the cohesive strength of the lattice and fracture occurs. The creep phenomenon is thought to occur by a similar mechanism although glide is spread over more planes and distortion accumulates along the grain boundaries, leading to the formation and coalescence of cavities there. Glide also occurs in fatigue stressing, and leads to extrusion of material from the slip bands with resulting formation of cavities and cracks.

2.4.2 *The Dislocation*

The simplest mechanism by which slipping could occur would be by the rigid body translation of the material on one side of the slip plane relative to that on the other. The atoms are not rigidly connected together, however, and such a movement would in any case require much more energy than is found to be necessary. The atoms slip consecutively, not simultaneously, and it is possible to draw a line in the slip plane separating a slipped from an unslipped region. This boundary is called a *dislocation line,* which must always either form a closed loop within the crystal or end at the free surfaces of the crystal.

In real crystals imperfections exist, and these introduce a concentration of stress from which the movement of a dislocation line may be initiated. The amount and direction of the relative movement between the atoms which are displaced can be described by a vector, which is known as the *Burgers vector* after its proposer.

There are several possible configurations which can be

described as dislocations, and they can be classified according
to the relative direction of the dislocation line and the Burgers
vector.   If these are normal then the dislocation is termed an
edge dislocation.   If they are parallel it is termed a *screw
dislocation*.   Both types are illustrated in Fig. 2.24 which
gives, of course, a highly idealized picture of the atomic
lattice.   In these illustrations it will be seen that the
magnitude of the Burgers vector is equal to one lattice spacing
– this is a dislocation of *unit strength*.   It will be noticed that
the dislocation line sweeps through the crystal in the direction
of the Burgers vector in the case of the edge dislocation, but
normal to it in the case of the screw dislocation.   Considering
the edge dislocation in Fig. 2.24 we see that the atoms are
squeezed together in the upper part of the lattice and spread
out below.

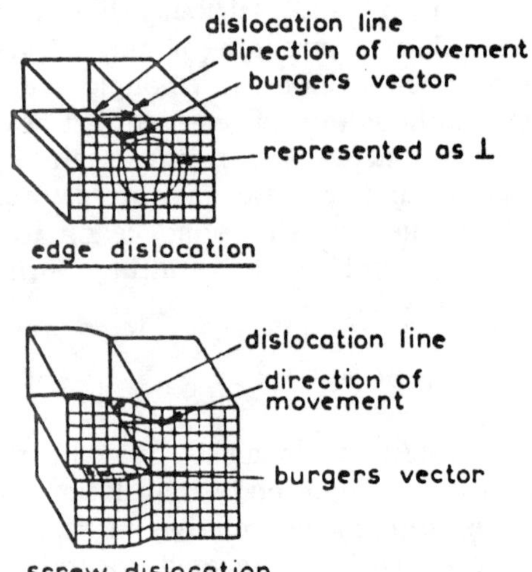

Fig. 2.24 Illustrations of dislocations

This 'transverse cross-section' of the dislocation line
associated with the edge dislocation is often denoted by the
symbol ⊥ .   It is easily realized that edge dislocations of the
same sign ⊥ ⊥ will be repelled from one another, whereas
those of the opposite sign ⊥ ⊤ will be attracted and tend to
cancel or annihilate one another if in the same plane.

The concept of dislocations was introduced, independently,
by G.I. Taylor, E. Orowan, and M. Polanyi, in 1934, and
considerable progress in the understanding of the physics of
metallurgy has been made since then by many distinguished
scientists, notably N.F. Mott and A.H. Cottrell in Cambridge

University, F.C. FRANK, and B.A. BILBY.   The historical development of the subject can be traced by reference to the book by COTTRELL[3].   For the present purpose it is sufficient to realize that most, if not all, of the phenomena which occur both in the deformation and crystallization of metals can be explained in terms of imperfections which can exist in the atomic lattice.   For many years it was difficult to understand why metals deformed at such relatively low stresses in comparison with the stresses required to disturb the perfect lattice.   The concept of the dislocation has removed this barrier to understanding, and resulted in significant practical advances in metallurgy.   With the advent of the electron microscope, it is now possible actually to observe the traces of dislocation lines and thereby verify the way in which they move and interact.

There are usually several possible slip systems in any crystal, and practical metals are polycrystalline, so that the movement and interaction of dislocations can be exceedingly complex.   Herein lies the reason for the necessity of the two branches of metallurgy — engineering and physical.   Although at first sight it might seem possible to be able to predict the macroscopic behaviour in terms of stress and strain from a knowledge of interatomic forces and the movement of dislocations, in practice the difficulties associated with such an approach are insurmountable, and recourse must be made to a 'macroscopic' theory of plasticity.   This theory has for its starting point the assumption of an idealized material whose properties are considered to be completely definable in terms of a basic stress-strain curve.   By using such a theory, a limited amount of information may be obtained about the way a metal will behave when subjected to certain stress systems; to understand the effects of metallurgical treatments it is helpful to consider the likely behaviour in terms of the movement of dislocations.

Complicated though the interaction of dislocations may be, it is possible to discuss a few important situations which arise. Often, for example, the interaction of dislocations results in the ends of a dislocation line being effectively 'locked' within a crystal by neighbouring dislocations moving on intersecting planes.   If the plane on which the dislocation is locked is also the most favourably orientated glide plane, then a special dislocation movement occurs upon application of a stress.   The locked dislocation line is referred to as a Frank-Read source, and its progress during stressing is indicated in Fig. 2.25, A and B being the initially locked ends of the dislocation.

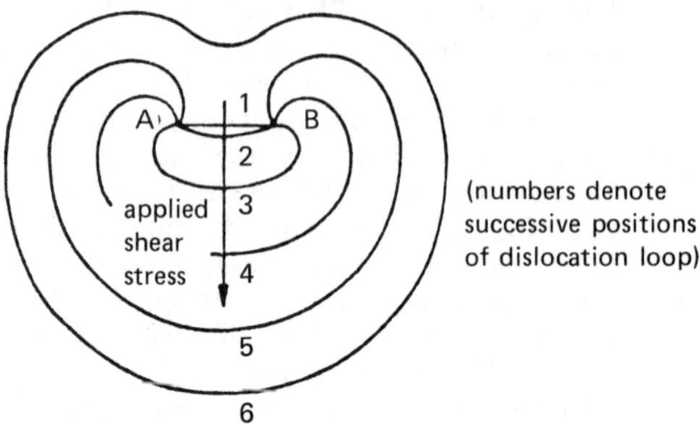

Fig. 2.25 The Frank-Read source

The loop grows by sweeping around A and B until the two sides meet forming a complete loop *and another line dislocation between* A *and* B, which can then repeat the process. Under the applied stress the loop moves out to the boundaries of the crystal, being followed by a succession of similar loops, a process which continues for as long as is necessary for dissipation of the applied energy. A mechanism of this type is necessary to provide the large number of dislocations observed in the slip bands.

Not only in the explanation of deformation behaviour has the dislocation become essential − it has also been shown that the *rapid* growth of crystals from vapour or solution is not possible unless the crystal has steps on its faces to receive deposited material. Such steps are also provided by dislocations, which can provide the energy distribution that favours deposition, and it may be seen from Fig. 2.24, for example, that the screw dislocation is particularly suited to crystal growth. Deposition of atoms occurs along the step, and growth proceeds by a spiral or screw rotation of the step round the point of emergence of the (stationary) dislocation line. Patterns of growth of this type have been observed with the electron microscope.

Another important type of lattice irregularity is the *vacancy*, or the hole which exists due to the presence of a vacant atomic site. Although vacancies can move throughout the structure by a process of diffusion, such internal movements produce no change at the surface of the crystal which is therefore not deformed. Vacancies tend to diffuse or *condense* together, thereby forming larger vacancies. When a sheet of vacancies becomes large enough it effectively collapses in the middle forming a dislocation barrier which cannot move or *is sessile,* as illustrated in cross-section in Fig. 2.26.   Such sessile dislocations form effective barriers to the passage of other dislocations.

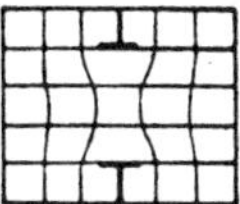

Fig. 2.26 The 'sessile' dislocation

2.4.3   *Hardening, Recrystallization, and Recovery*

Having obtained an over-all picture of how plastic deformation proceeds on the sub-microscopic scale, it is now possible to consider how metals may be made more resistant to plastic flow, i.e. hardened.   It seems clear that any barrier to the movement of dislocations will be effective in hardening the metal.   A convenient division of the possible barriers to dislocation movement is as follows:-

(1)   Grain boundaries (and sub-boundaries)
(2)   Other dislocations
(3)   Solute atoms
(4)   Precipitated phase.

Considering the first of these, in a polycrystalline metal, it is unlikely that a slip system operating in one crystal will be parallel to one in a neighbouring crystal, unless the metal is one having a large number of alternative slip systems (e.g. α ferrite).   Thus grain boundaries are effective obstacles and cause the piling-up of dislocations on the active slip planes, as sketched in Fig. 2.27.

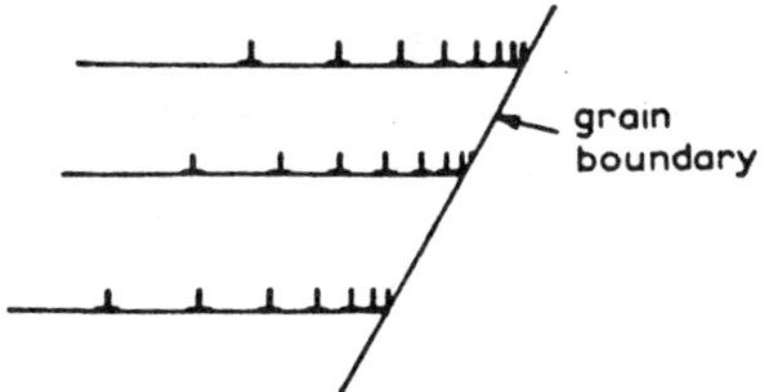

Fig. 2.27 A pile-up of dislocations

Comparing a polycrystalline metal having an h.c.p. lattice (e.g. zinc) with one having f.c.c. (e.g. aluminium), it is found that zinc attains a higher yield stress.   This may be attributed to the scarcity of available slip planes in the hexagonal zinc structure, and it is much easier for the relief of grain boundary stresses due to piled up dislocations to occur in the f.c.c. aluminium, since there are four possible planes available for slip in each crystal. This behaviour is even more striking when it is realized that the basic shear stress/shear strain curve for a single crystal of aluminium is very much higher than that for zinc, so that the hardening effect of grain boundaries must be considerable. It is also evident that the smaller the grain size the greater will be the area of boundary present and therefore the greater the strength (in fact it is proportional to the area of the boundary). Other obstacles which are similar to grain boundaries and may be included in this first category are lamellar eutectic structures, surface oxide films on single crystals, and sheets of stacking faults. (*A stacking fault* is a zone of the lattice which has slightly changed due to some external disturbance, e.g. phase transformation or cold work; a typical example is when dislocation slip on a close packed plane changes the f.c.c. structure into h.c.p.)

The second barrier to the movement of dislocations is their interaction with dislocations moving on other planes, and is especially important in discussing the phenomena of work hardening and recovery. Many types of interaction are possible, some causing annihilation, others producing sessile dislocations, for example. The most effective are those which produce steps in the dislocation lines themselves, the steps being known as *dislocation jogs*. Such jogs can be caused by intersecting edge dislocations, intersecting screw dislocations, or by intersecting edge and screw dislocations. By way of illustration, the simplest type of intersection is shown in Fig. 2.28. The continued interaction of moving dislocations forms trails of 'jogs' or lattice imperfections, which requires additional energy leading to work hardening of the structure. The interaction of screw dislocations is interesting as it leads to the movement of jogs generating either lines of vacant lattice sites or lines of atoms moved from normal sites into interstitial positions. The effect of reversal of stressing is important, and it should be possible to explain, at least qualitatively, many of the phenomena observed in conventional and low cycle fatigue, such as the Bauschinger effect and work softening, in terms of the interaction of dislocations and jogs. Support for this explanation of work hardening is obtained from the observed fact that single crystals of the hexagonal metals, which have no possible intersecting slip planes, exhibit little or no work hardening at all.

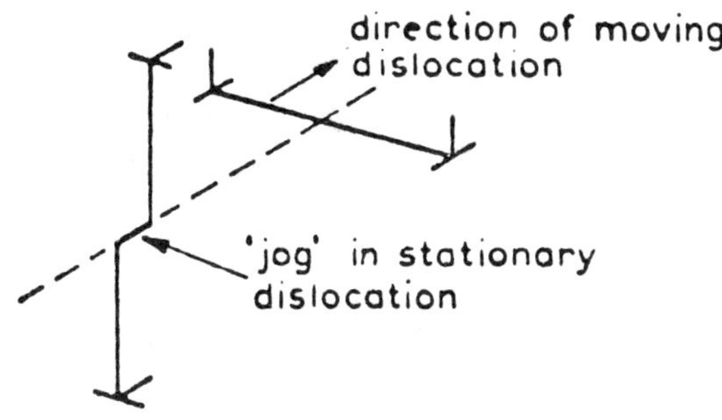

Fig. 2.28 Interaction between dislocations (from COTTRELL[3],
by courtesy of Edward Arnold.)

On the other hand, single crystals of cubic metals work harden considerably, although if special precautions are taken to restrict slip to one plane only, work hardening can be completely suppressed (a phenomenon known as *easy glide*). In polycrystalline metals, of course, a certain amount of hardening is caused by the piling-up of dislocations at the grain boundaries, so that hexagonal metals do exhibit a certain amount of work hardening.

A third barrier to the movement of dislocations is that provided by the presence of solute atoms. There is generally a difference in size between the atoms of the parent lattice and the foreign solute atoms, so that the lattice of the substitutional solid solution is strained or distorted. Because of this, more energy is required to move the dislocation through, and the alloy is thereby hardened. The effect of hardening by solute atoms is illustrated for copper-nickel alloys in Fig. 2.14, from which it will be evident that the maximum effect is obtained, as would be expected, when the metals are present in approximately equal amounts. The effect is even more marked for single crystals of silver-gold alloys; although there is less than 0.2 per cent difference in the atomic radii, the 50 per cent alloy is about seven times harder than pure gold (the stronger pure metal). It would be expected that the greater the difference between the atomic diameters the greater would be the barrier and, therefore, the hardening effect. Although this is broadly true there are discrepancies between the observed hardening and that expected from the theories developed on this basis, and it appears that the relative valency of parent and solute atoms has an effect. The bigger the difference in valency, the greater the hardening effect. There is also evidence that a static dislocation in a substitutional solid solution may change its character, since there will be a tendency for the smaller and larger atoms above and below a dislocation to migrate by diffusion into regions of lower energy. Thus the smaller atoms will migrate to the 'squeezed-up' side of the dislocation and an *atmosphere* of solute atoms will collect round the dislocation, effectively locking it. When the solute atoms are present within the interstices of the lattice, as in interstitial solid solutions, they tend to diffuse into the cavity immediately under the dislocation line, again locking it. An extremely important interstitial solid solution is that of carbon in iron. The carbon atoms occupy the interstices of either the γ-iron f.c.c. structure, or the α-iron b.c.c. structure, there being much less room in the latter. As previously mentioned, quenching of the γ-iron produces a b.c.c. structure distorted by the interstitial carbon atoms, because they have insufficient time to diffuse and form the cementite phase. This structure is so distorted by the interstitial carbon atoms that dislocations cannot pass through it, so martensite is exceedingly hard and brittle.

The fourth type of barrier which may be provided against the movement of dislocations is the *precipitate*. This is usually formed from a supersaturated solid solution which is itself relatively hard due to the distortion of the lattice caused by the solute atoms, which are usually separated by a distance of only about three atomic spacings. During precipitation the distance between the precipitated particles increases, as well as the size of the particles themselves, and it appears that a maximum hardening is attained when the dispersion is between 25–50 atomic spacings. Further precipitation results in over-ageing, and the particles can eventually be seen under an optical microscope {the dispersion being > 1000 Å (10^{-4} mm or about 4 micro-in.)}, by which time the hardness has considerably decreased.

Having discussed the purely mechanical behaviour of the atomic lattice when subjected to stress, particularly the role of the dislocation and other defects, next consider the effect of thermal treatments. The processes of *recrystallization* and *recovery* are distinguished by the fact that softening without recrystallization occurs during *recovery*. During *recrystallization* a metal which has been cold-worked is heated until new crystals grow from nuclei to replace the strained ones, the new lattice orientation being different from the old. Such growth is termed *incoherent* and proceeds by the *migration* of the large-angle boundaries of the new crystals. When all the strained material has been replaced by these new crystals recrystallization is complete but the crystals will continue to grow by consuming each other (*grain growth*).

Recrystallization is a thermally activated process and the nucleation follows a well-known law which applies to many such processes, examples being the self-diffusion of atoms or grain boundaries, grain growth, and creep. This law is derived from statistical mechanics and is of the form:-

$$r = A \exp\left[-\frac{Q}{RT}\right] \qquad (2.12)$$

where

r = the rate at which the process occurs (or the number of nuclei formed per unit time),
Q = the activation energy for the particular process concerned,
R = the universal gas constant,
T = the absolute temperature,
A = a constant.

This is *Fick's law* of diffusion rate, the equation actually being accredited to ARRHENIUS, and expresses the fact that for a reaction which requires an activation energy of Q (usually expressed in units of cal/g atom or cal/g mol), the number of atoms with sufficient energy to overcome the potential energy of the barrier will be proportional to $\exp(-Q/RT)$. It is often found in an experiment that, by plotting a graph of experimental values of log r (or ln r) against $1/T$, a straight line results and the slope gives the value of $-Q/R$, from which Q can be estimated (care being taken to cater for the actual logarithmic base used). Many chemical reactions have activation energies of about 10 kcal/g mol. The activation energy for the nucleation of new crystals is much higher and very similar to that for grain growth, lying between 50 and 70 kcal/g mol. Recrystallization is also a function of the annealing time t, the rate of formation of nuclei being very slow at first, then rising to a maximum after which it gradually decreases to a very small value.

There is evidence that recrystallization is nucleated at sites where turbulent slip has occurred, for example in deformation bands or grain boundaries where the crystal lattice has become locally distorted and curved by the passage or accumulation of dislocations. It has been shown that single crystals deformed in laminar slip (easy glide) will not recrystallize. Both recrystallization and subsequent grain growth are dependent on the mobility of grain boundaries, which in turn depends both on the change in lattice orientation across the boundary and on the orientation of the boundary itself. The most mobile boundaries are either those with very small angular changes in the lattice, made up of symmetrical edge dislocations, or those which are incoherent with large angular lattice change. The latter type of growth is important in producing a recrystallization texture quite different from the original worked structure.

Recovery is thought to be due to the running together and mutual annihilation of dislocations, the energy for this having been supplied by thermal activation. For this to happen it may often be necessary for dislocations to *climb* out of their slip planes by the migration or diffusion of vacancies, or for the diffusion of atoms to provide the necessary half sheet of atoms to annihilate a dislocation. Dislocation climb often continues until the dislocations form 'vertical' walls one above the other as indicated in Fig. 2.29. Such walls of

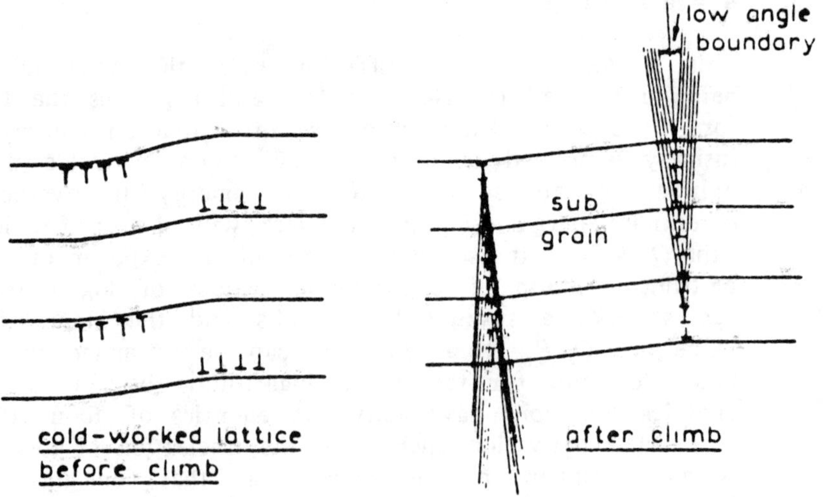

Fig. 2.29 Dislocation climb leading to 'polygenization'

dislocations effectively constitute low angle grain boundaries, so that a sub-structure is formed within the main crystal system. This process is known as *polygenization,* and leads to the formation of sub-grains with softening of the metal due to the uniform spreading out of the dislocations throughout the structure.

2.4.4 *Creep, Sharp Yield Point, and Strain-Ageing*

As described in the previous section, solute atoms can diffuse towards a stationary dislocation and form an atmosphere around it. Particularly important cases are those of carbon and nitrogen in body-centred iron. These elements can occupy interstitial positions at the face and edge centres of the cubic cell leading to a tetragonal distortion of the lattice. This enables the solute atoms to interact with both edge and screw dislocations, so that each dislocation gathers around itself an atmosphere of solute atoms which are attracted towards the dislocation.

When an external stress is applied to such a system, the dislocation will try to move in the normal way, but will be held back by its atmosphere of solute atoms, which provides an anchoring force. The dislocation will only move if the external force is raised to a sufficiently high level to overcome this anchoring force, unless the dislocation can move slowly enough to drag its atmosphere along with it. Since this latter process can be effected by diffusion of the solute atoms, it could occur at a lower applied stress, albeit at a very slow rate. This type of dislocation movement is undoubtedly important in creep processes, although the mechanisms occurring in creep are numerous and complex. In the primary stage of creep there is movement of dislocations much as in high-speed plastic flow, but during the secondary phase the important mechanism seems to be that of dislocation climb. A balance is struck between the hardening due to the turbulent movement of dislocations and a softening or recovery due to continual removal of dislocations from the interior of the sub-grains to their boundaries. There comes a time when the dislocation density and therefore the resistance to deformation remains constant, and this is the secondary phase. Distortion is so concentrated at sub-grain boundaries due to the climb of dislocations, in fact, that voids nucleate there and coalesce during the tertiary stage, and fracture occurs after a relatively small *over-all* distortion. Since vacancy diffusion occurs more rapidly the smaller the grain size, and there is a large area of grain boundary associated therewith, both dislocation climb and void formation and coalescence are encouraged by small grain size, resulting in greater creep. Thus it would appear that coarse-grained material is desirable for high creep resistance, which is rather different from the usual requirement for metals to have a fine grain size.

In general, metals obey Andrade's creep law in the primary stage, namely:-

$$\epsilon = \alpha t^{1/3} \qquad\qquad (2.13)$$

where ϵ = creep strain,
      t = time, and
      α = a constant at a given stress and temperature.

To allow for the steady state secondary phase it is reasonable to add a term βt to Equation 2.13, and Graham and Walles have suggested adding a term γt^3 to incorporate the tertiary phase, so that a useful constitutive equation for creep including all creep phases is:-

$$\epsilon = \alpha t^{1/3} + \beta t + \gamma t^3 \qquad\qquad (2.14)$$

It is found that the creep rate in the steady-state secondary phase fits Fick's law very well. The activation energy for the process is, as would be expected, a decreasing function of the stress, and Cottrell determined values of about 20 kcal/g mol for zinc crystals.

Usually, the rate of straining is too high or the temperature too low for the creep process just described to take place. The dislocations are anchored along their entire length by immobile atmospheres and must be pulled away by an applied force exceeding the anchoring force. Once they have been pulled away the force required to keep them in motion is much lower, of the same order as exists in the absence of atmospheres. The material can be either in an unyielded or *strain-aged* condition, or in an overstrained condition in which the dislocations are free to move and cause plastic flow. Cottrell has proposed this as the explanation of the phenomenon of the sharp yield point which exists in some metals containing certain impurities. The most important examples are soft iron or mild steel, the optimum amount of carbon being between 0.001 and 0.1 per cent. With higher carbon contents or with other elements present the effect is less apparent, except at low temperatures, presumably due to the presence of other obstacles (such as cementite) which require higher stresses to remove the dislocations than those required to pull them from their atmospheres. (At low temperatures the anchoring force of the atmosphere of solute atoms is raised due to their lower thermal energy.)   Yield points of this type have been observed in molybdenum, cadmium, zinc and brass containing nitrogen as an impurity.

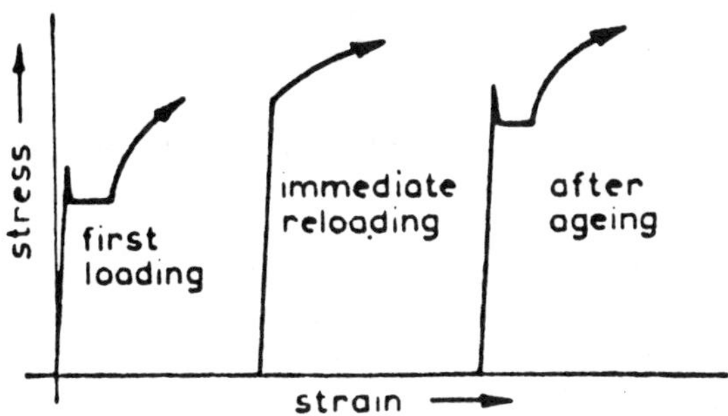

Fig. 2.30 Illustrating 'sharp' yield point

This effect is illustrated in Fig. 2.30, the salient features being the existence of an upper yield point associated with the freeing of the dislocations from their atmospheres followed by a period of yield at low stress; the complete absence of a sharp yield point on immediate reloading due to the fact that the dislocations have escaped from their anchoring atmospheres; the return of a higher sharp yield point when the specimen is left for some time (accelerated by ageing at temperature) due to the diffusion of the atmospheres of solute atoms back to an even greater number of dislocations in the overstrained metal.

This phenomenon is of great importance in practice, and presents one of the biggest problems in the processing of mild steel sheet.   The presence of the sharp yield point leads to the formation of *Lüders bands* or *stretcher strains*, i.e. markings on the sheet metal when subjected to deep drawing or pressing operations.   To mitigate this effect the sheet is *temper rolled* (i.e. by giving a few per cent reduction in section), to remove the sharp yield point, but it must be used within a short period (e.g. a day or two at room temperature), before the sharp yield returns due to ageing.   The discontinuous sharp yield, in which a 'break-away' stress is involved, leads to the initiation of yielding at certain discrete points of stress concentration within the material.   Such stress concentrations can exist at gripping points, surface imperfections, or at crystal boundaries, so that yielding starts at these points and gradually spreads through the material until the whole piece has yielded.   *Blue-brittleness* of mild steel occurring at a temperature of about 300°C can also be explained by the rapid formation of atmospheres which require a sharp yield point to break away the dislocations, leading to a succession of yield points on the stress–strain curve. The phenomenon is therefore strongly rate-dependent. A similar effect is observed in cold worked aluminium alloys containing copper.

2.5 General Relations between Metallurgy, Deformation, and Properties

It will be appreciated from the preceding discussion that the mechanical properties of metals and alloys are dependent on many factors. For example, they depend on the proportions of the constituent metals in an alloy system, on the types of phases which exist, on the heat treatments which have been received by the alloy, and on the amount of plastic deformation which has occurred. The interaction of these latter factors broadly determines whether a metal is being hot worked or cold worked. In the final section of this chapter an attempt is made to summarize the more important considerations relating to the interplay between such treatments and material properties.

2.5.1 *The General Effects of Metallurgical Composition and Heat Treatment on Final Properties*

An example of the general effects of metallurgical composition and heat treatment on final properties has already been discussed in connection with Fig. 2.14, for substitutional solid solutions. It is generally true that the presence of solute atoms creates a straining of the lattice which provides barriers to the movement of dislocations with a resultant hardening of the (cold) alloy, the maximum effect being obtained at about 50 per cent concentrations. The ductility exhibits the reverse trend, indicated in Fig. 2.14.

On the other hand a binary alloy comprising two metals which are insoluble in each other in the solid state, or nearly so, exhibits an approximately linear variation of hardness between those of the pure constituents. Often, however, the eutectic is stronger than would be expected on this simple basis due mainly to the large number of phase boundaries which provide dislocation barriers.

In many alloys, the metals exhibit a limited amount of solid solubility in one another, so that the variation of hardness and ductility of such alloys exhibits a cumulative effect comprising the two types of behaviour just described. There are also many binary alloys which possess a composition range or ranges over which intermediate phases or intermetallic compounds are formed such as, for example, cementite in the iron carbon system.   Such intermediate phases are generally extremely brittle and hard at normal temperatures, although there are a few, such as the beta phase in brass at high temperature, which may be comparatively ductile.   The properties of binary alloys which form intermediate phases can be derived by regarding the intermediate phase as a third component common to each of two 'pseudo binary' systems. In binary systems which form more than one intermediate phase it is generally found that mixtures of the intermediate phases are extremely brittle.

The effect of alloy composition on the rate of strain hardening or the slope of the stress-strain curve is complex, but certain generalizations can be made.   Heterogeneous alloys (i.e. those which do not exhibit solid solubility) possess (true) stress-strain curves which lie broadly parallel to those of the pure metal or solid solution comprising their main substance. Ductility (as measured, say, by elongation to fracture in the tensile test) decreases as hardness increases.   Solid solutions, on the other hand, may show either higher or lower strain-hardening rates.

The main heat treatments which can be given are those of annealing type and those of precipitation type.   By and large, annealing and normalizing treatments result in diffusion and hence in recovery and recrystallization of the structure of the alloy, with an attendant reduction in hardness and increased ductility.   A previous history of cold work may thereby be wiped out, its only effect having been to accelerate the annealing process by nucleation of sites for grain growth. In general, the smaller the previous reduction, the higher the temperature necessary to effect recrystallization, and the larger the resulting grain size.   There is a critical range of reduction which results in excessively large grains on annealing, and must therefore be avoided.   For low carbon steels, the range is quite large, being from 5 to 20 per cent, whilst for certain aluminium alloys there may be two critical ranges, one at low values of strain (2-5 per cent), the other at high values (80-90 per cent compressive strain).   The effect may be offset to some extent by keeping the annealing temperature as low as possible, and the annealing time as short as possible.   If the temperature greatly exceeds the minimum necessary for recrystallization the grain size will become large due to grain growth, even after large reductions. Many metal forming operations lead to inhomogeneous deformation occurring throughout the volume of the workpiece, which produces similar effects particularly in aluminium, titanium and nickel alloys.

Precipitation is the most common type of phase change occurring. It leads to a hardening of the alloy due to the barriers to dislocations created by the precipitated particles. It is always preceded by some form of solution treatment and quench, in order to produce the necessary unstable super-saturated solid solution. The quenching and tempering of steels may be regarded as a metallurgical treatment of this type. The effects of temperature and time of ageing are quite pronounced, and have been studied for most of the important alloys.

2.5.2 *The Interrelation Between Hot Working, Grain Structure, and Properties*

It is difficult to generalize about the subject of this section, since the interaction between hot working and properties is extremely complex. In the first place it is necessary to distinguish between hot and cold working, a task which many engineers seem to have difficulty in understanding. Perhaps the best definitions have been given by JOHNSON and MELLOR[11], abbreviated somewhat as follows:-

(i) **Hot working:-** plastic deformation of metal at high homologous temperature and at a rate of straining low enough to avoid significant strain hardening and to allow recovery and/or recrystallization to occur without excessive grain growth.

(ii) **Cold working:-** plastic deformation of metal at low homologous temperature and at a rate of straining high enough to avoid significant recovery and/or recrystallization and to allow moderate strain hardening to occur, without fracture.

The main reasons for hot working are economic, in the sense that the easiest way traditionally to produce large tonnages of prismatic section or large numbers of forged or extruded pieces is to cast large ingots and then deform them by forging, extrusion or rolling whilst in the soft state associated with high homologous temperature. Again, as pointed out in the last chapter, there are arguments for thinking that the most efficient way to deform metals is by making them approximate as nearly as is reasonably possible to the molten state, although this could be dangerous in some cases.   Also, since metals are generally quite expensive, especially those required for modern technologically advanced artefacts, such as in aerospace applications, there is a great incentive to reduce machining of parts to a minimum and even to eliminate completely some of the massive bulk machining of parts requiring the strength of the wrought metal, which has been necessary in the development of modern technology.   Finishing or near-finishing to size by cold working is desirable anyway to confer both dimensional accuracy and good surface finish and enhanced fatigue strength.   In modern terminology, there is a great incentive for Industry to develop processes of *Nett Shape Forming or Near Nett Shape Forming*.

Then again, a certain amount of hot working is very desirable.   The cast structure of any ingot or casting is highly directionally orientated, often comprising large columnar crystals especially near surface boundaries, and hot working breaks up and refines this generally undesirable and weak cast structure.   Diffusion of alloy constituents is facilitated and brittle films or particles of hard constituents are broken up and distributed more evenly throughout the volume.   Also, most hot-working processes are purposely made compressive so that cracks, cavities and blowholes are encouraged to close up and weld together.   As a matter of fact, these hot-working processes are themselves 'inspection' tests of metal ingots, since any excessively large flaws will reveal themselves during processing by the breaking up of the piece, eliminating it from further processing.

Some metals, such as lead and tin, have very low melting temperatures and may be hot worked at room temperature. The upper temperature of the hot working range is generally determined by the lowest melting point constituent of the alloy.   If that temperature is exceeded, the alloy will simply break up and in practice the temperature of working must be kept well below that, to allow for uncertainties in composition and for adiabatic or near adiabatic heating from the work done during the processing.

The rate of deformation is clearly one of the more important variables in hot working.   Deformation rates which are too high are undesirable since the hot-working process becomes a cold-working process.   Hammer forging by impact blows is often employed as being a convenient and inexpensive way of achieving high deformation forces on the workpiece. Tensile stresses must be avoided at all costs, however, because of the possibility of brittle failure and dangerous explosive disruption of the workpiece. For that reason hot hammer forging is usually confined to *partially closed dies*, to encourage all-round compressive stresses in the workpiece. (Completely closed dies are rarely used because of the possibility of overstressing the press and die components.) Actually, the distribution of stress in a piece struck by an impact blow is different from that consequent on slow deformation. Due to the inertia of the metal of the workpiece the stresses become localized at the regions of contact. Thus, if it is desired to work a large piece right through to the centre of its volume it should be squeezed slowly in a hydraulic press rather than in a hammer forge. In general, for these reasons, hot hammer forging should be reserved for parts having relatively small cross-sectional dimensions.

The effects of phase changes in the solid state on behaviour during hot working are very complex and it is impossible to generalize. One of the main reasons for this is that equilibrium diagrams really only define the behaviour of alloys after equilibrium has been reached and that always takes far longer than the time available, even in the slowest of commercially viable hot-working routines. Therefore, it is beyond the scope of this book to try to catalogue the individual idiosyncrasies of particular metallic alloy systems. In fact, such information is not easily found, constituting as it does the accumulated wisdom of countless industrial metallurgists and metallurgical engineers over the years. It is generally necessary for anyone contemplating using a particular alloy system to make an extensive literature survey before embarking on a manufacturing programme using that alloy.

2.5.3 *The Interrelation Between Cold Working, Grain Structure, and Properties*

The main effect of cold working a polycrystalline metallic alloy is to cause a sub-microscopic breaking up of the grain structure due to the interaction of moving dislocations, which also leads to a work hardening (or strain hardening) of the metal and a consequent loss of ductility. The grain structure of all metals is affected in a similar manner by cold working. If the grains are originally equiaxed they will each become distorted in a similar manner to the over-all distortion of the piece. Thus, if a billet of material is elongated 400 per cent by rolling, without spreading, it will have been reduced by 80 per cent, and each of the formerly equiaxed grains will be five times their original length and one-fifth of their original thickness.

The resultant properties of the metal can be obtained with reasonable accuracy from its basic stress-strain curve, derived from one or other of the various tests described in the previous chapter, by finding the appropriate amount of effective true strain suffered. The distortion of the crystal structure which occurs leads to some anisotropy in properties, and this must be borne in mind if great accuracy is required.

In production processing by cold working, there is a limit to the grain size which can be cold worked and give a satisfactory product. If the grain size is too large, so that the section being deformed contains only a few grains (say 50) then it will naturally flow irregularly. This is especially true of the hexagonal metals which have few available slip planes, such as the zinc and magnesium alloys. Free surfaces on the component being processed will develop an irregular outline resembling an orange peel, which may be unacceptable in service. (Sheet metal is tested for this defect as part of the Erichsen test, described in Chapter 4.)

Under certain unfavourable conditions the presence of large grains encourages cracking along grain boundaries during cold working. Such brittleness is found in tungsten and molybdenum, as well as in zinc and magnesium, and these metals show considerable increase in ductility when their grain size is small. (If the temperature of working is increased into the hot-working range, ductility returns even for coarse-grained metals.)

It should be pointed out here that, although it is possible to correlate different forms of cold-working processes like rolling, forging, simple compression or tension as far as yield properties are concerned (by using the basic true-stress/true-strain curve with the concept of equivalent strain), it is not possible so to correlate ductility. For example, although it is common practice to extend a piece by around 70 per cent in a single pass through a rolling mill, it cannot be extended by more than about 25 per cent in simple tension without necking and fracture. In general, compressive ductility is very much higher than tensile ductility and the strain to fracture is obviously a function of the applied hydrostatic pressure (as discussed at the end of the previous chapter), amongst other factors.

In addition to the difficulty of explaining the effects of cold work as far as ductility is concerned, the effects of cyclic stressing into the plastic range are not fully understood. It seems from research work that it is possible to bring a metal into a stable yield stress state by plastic cycling (i.e. cycling the stress state through the plastic range), the final stable yield stress being higher than that which would have been expected from the basic stress-strain curve, associated with the total equivalent strain imposed during cycling.

One important consideration in cold working is that, provided the tools and lubrication are suitable, accurate dimensions and good surface finish can be produced. Another is the great benefit of cold working introducing strain hardening, allowing the use of cheaper steels. It is these considerations which help to dictate the use of cold-working processes towards the end of the production cycle.

REFERENCES

(1) ROLLASON, E.C., *Metallurgy for Engineers.*
Edward Arnold, London (3rd Ed.), 1961.

(2) SMALLMAN. R.E., *Modern Physical Metallurgy.*
Butterworths, 1962.

(3) COTTRELL, A.H., *An Introduction to Metallurgy.*
Edward Arnold, 1967.

(4) CADDELL, R.M., *Deformation and Fracture of Solids.*
Prentice-Hall, 1980.

(5) VAN VLACK, L.H., *Elements of Materials Science.*
Addison-Wesley, Reading, Mass., 1959.

(6) PASCOE, K.J., *An Introduction to the Properties of Engineering Materials.* Blackie, Glasgow, 1961.

(7) BALL, J.G., "Teaching of Metallurgy to Engineers",
Proc.I.Mech.E. 1961,175,169.

(8) DIETER, G.E., *Mechanical Metallurgy.*
McGraw-Hill, New York, 1961.

(9) CRANE, F.A.A. and CHARLES, J.A., *Selection and Use of Engineering Materials.* Butterworths, 1984.

(10) ANDERSON, J.C., LEAVER, K.D., RAWLINGS, R.D. and ALEXANDER, J.M., *Materials Science.*
Van Nostrand Reinhold, (3rd Ed.), 1985.

(11) JOHNSON, W. and MELLOR, P.B., *Plasticity for Mechanical Engineers.* Van Nostrand, London, 1962.

Chapter 3

POLYMERS, ELASTOMERS AND REFRACTORY MATERIALS

3.1 Introduction

Natural materials such as wood and stone have been known and used throughout the history of man, but industry has depended much more upon the superior manufacturing properties and versatility of metals, especially iron and steel. In relatively recent times the range of available materials has been greatly extended, supplementing and often replacing metallic alloys for both domestic and industrial purposes. The most widely used of the new synthetic materials are the polymers commonly known as plastics, though these are in fact far simpler and more primitive materials than the natural polymers of muscles, ligaments and bones, from which much is still to be learnt[1]. There are also completely new demands from aerospace components and structures that require resistance to extremes of temperature and environment, met by specialized ceramic materials[2] related to but far transcending common pottery.

3.2 Polymeric Materials

Although inorganic polymers exist, based on silicon, these are of minor manufacturing importance in comparison with the organic polymers.

All the common plastics are essentially large molecules based ·on the four covalent bonds of carbon atoms[3]. The simplest is polyethylene or polythene, derived from ethylene, C_2H_4. It is important that this molecule contains a double bond between the carbon atoms, whose unsaturated nature allows one of the bonds to be transferred to another carbon atom, as shown in the following diagram:-

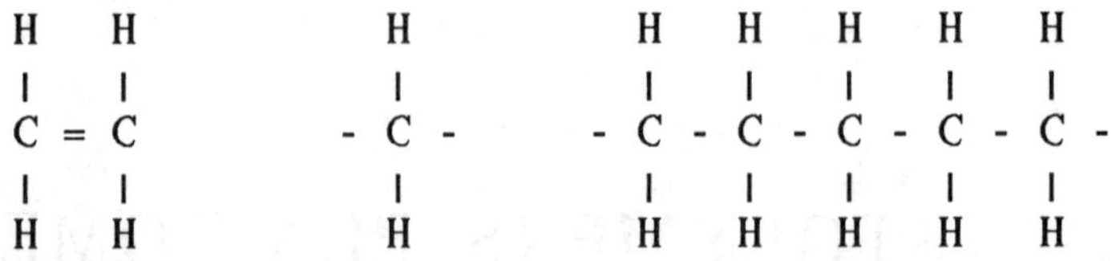

```
     H   H                 H            H   H   H   H   H
     |   |                 |            |   |   |   |   |
     C = C              -  C  -       - C - C - C - C - C -
     |   |                 |            |   |   |   |   |
     H   H                 H            H   H   H   H   H
   monomer               mer               polymer
  (ethylene)                            (polyethylene)
```

The basic repeating unit is known as the mer.   In most of the commercially important plastics the multiple n is very large, giving molecular weights based on C (=12) and H (=1) that may be tens or hundreds of thousands.

The mer may contain atoms other than hydrogen; for example PVC is based on vinyl chloride:

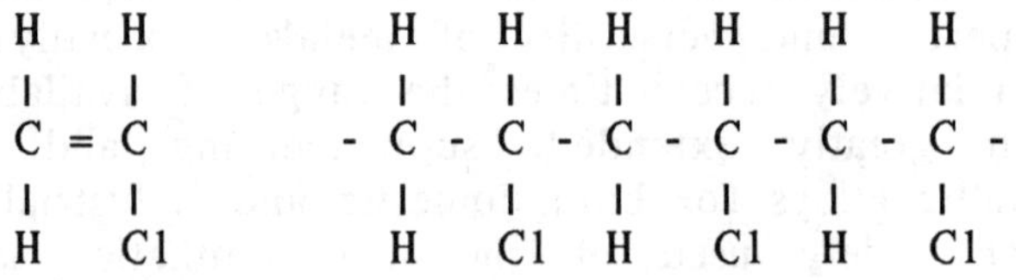

```
     H   H             H   H   H   H   H   H
     |   |             |   |   |   |   |   |
     C = C           - C - C - C - C - C - C -
     |   |             |   |   |   |   |   |
     H   Cl            H   Cl  H   Cl  H   Cl

   vinyl chloride        polyvinyl chloride
```

and PTFE contains no hydrogen, as can be seen from the diagram:-

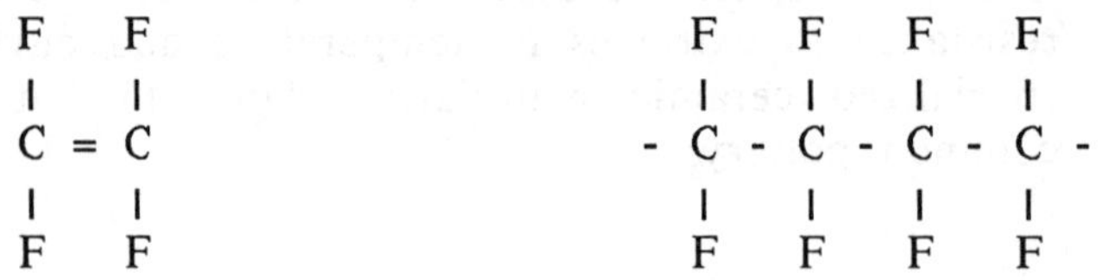

```
        F   F                      F   F   F   F
        |   |                      |   |   |   |
        C = C                    - C - C - C - C -
        |   |                      |   |   |   |
        F   F                      F   F   F   F

   tetrafluoroethylene          polytetrafluoroethylene
```

This well-known polymer has remarkably good anti-stick and lubricating properties, mainly because the large F atom shields the C forces between the molecules.

The maximum theoretical strength of such polymers is the covalent C-C bond (of the order of 300 kJ/g mol) but in fact the long molecules are highly intertwined and convoluted. The measured deformation strength is perhaps one hundred times lower, because the molecules can slide easily over one another, being held by cohesive bonds of low strength due to residual or van der Waals forces.

Longer chain molecules show greater interaction, especially when mutually aligned. High molecular weight polymers are thus stronger than their shorter analogues, and the strength increases with packing density; especially in the *crystalline* regions of close alignment, contrasted with more random *amorphous* regions. High density polyethylene (HDPE) for example, is harder and more wear resistant than the common material polythene.

The polymer chains may also be branched or cross linked:

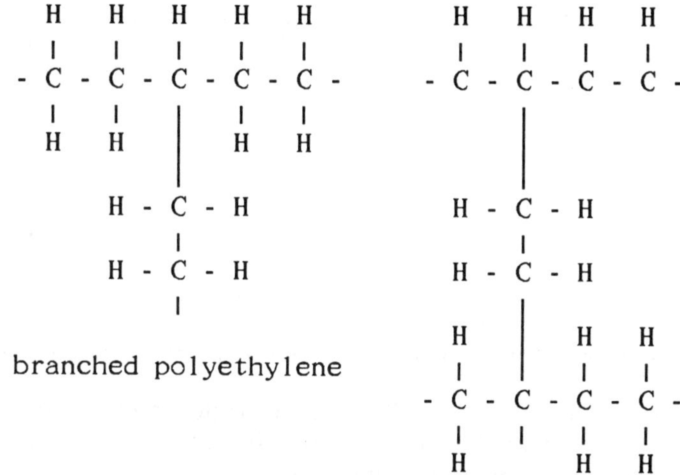

These are stiffer and stronger. Cross-linked polyethylene, produced by irradiation for example, will withstand boiling water, which softens the linear material. This can be valuable in sterilizing bottles and other containers.

The mer may also contain a benzene ring, as for example in polystyrene:

This again stiffens the molecule. A further stage of mer complexity is reached in phenol formaldehyde (bakelite) formed by a condensation reaction between phenol and formaldehyde with the elimination of water, catalysed by meta-cresol:

formaldehyde + phenol → → → → → → phenol formaldehyde + H_2O

This is a well known rigid and brittle *thermosetting plastic*, which forms a strongly interlinked network of atoms. The molecules are themselves naturally cross-linked.

Other polymers may be cross-linked by a second molecule, for example polystyrene linked by di-vinyl benzene:

di-vinyl benzene

3.3 Deformation Characteristics of Polymers

3.3.1 *Simple Models*

At high temperatures the shorter linear polymers behave as high viscosity liquids, or *melts*.

More commonly the response to a shear stress is not linear as in Newtonian fluids, but the shear strain rate is proportional to some power n of the shear stress τ (Fig. 3.1):-

$$\frac{d\gamma}{dt} = K\tau^{n} \qquad (3.1)$$

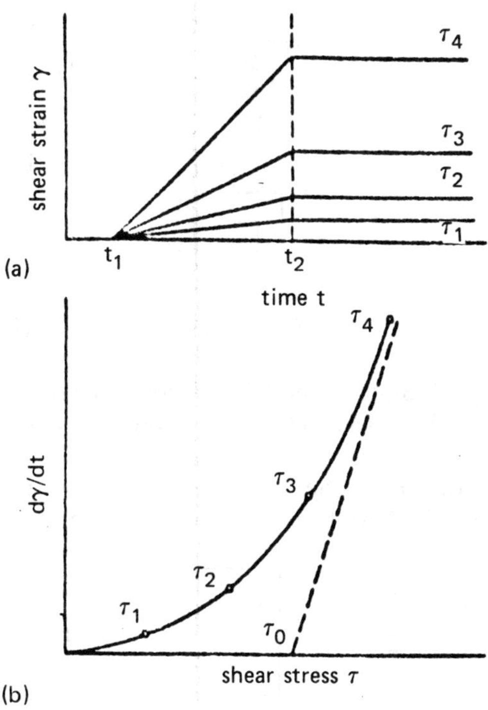

Fig. 3.1 The strain rate/shear stress relationship
for a polymer, and the Bingham model

It is often possible to approximate to this curve by a linear relationship:-

$$\frac{d\gamma}{dt} = \frac{\tau - \tau_{0}}{\eta} \qquad (3.2)$$

The materials following this type of behaviour are known as Bingham solids[4]. It is sometimes suggested that τ_0 represents a yield stress, such as may be found in some metals, but the linearization is simply a theoretical convenience and no real region of zero flow rate exists. Nor is the high shear stress response strictly Newtonian as the simple equation suggests.

The deformation of polymers is better described in terms of a viscoelastic model. Two elementary forms have been proposed, based on an elastic spring and a pure viscous dashpot in series or in parallel, as shown in Fig. 3.2. Electrical resistance/capacitance analogues can equally be used.

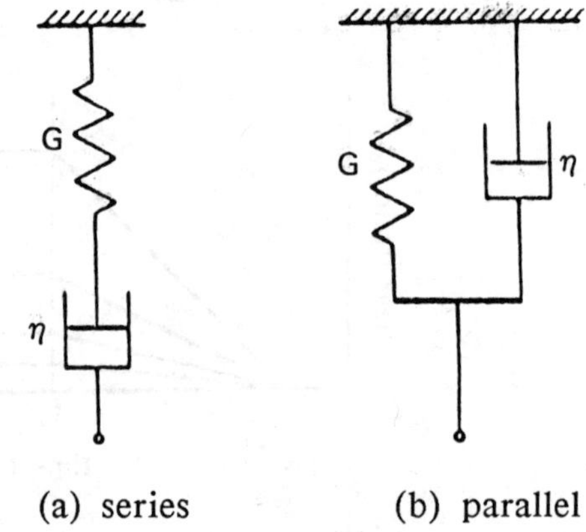

Fig. 3.2 Simple models of viscoelastic flow suggested
by MAXWELL (a) and by VOIGT (b)

(a) For the series model the response to a force suddenly applied and then released is shown in Fig. 3.3.

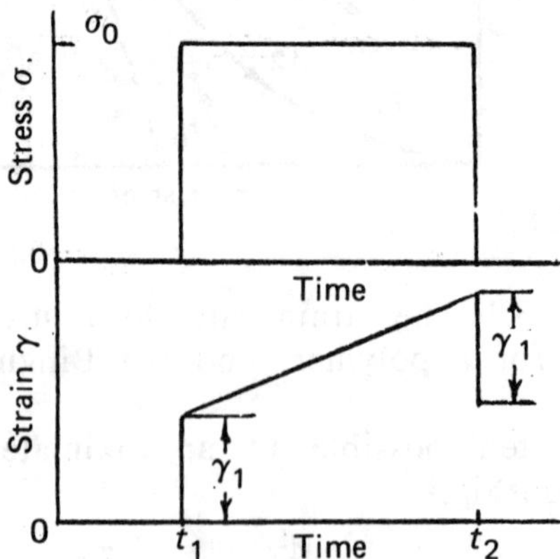

Fig. 3.3 The response of a MAXWELL model

At time t_1 when the load is applied, the spring responds instantaneously to produce a strain γ_1. The stress σ_0 then remains unchanged while the dashpot extends linearly with time until the force is removed at time t_2. The spring finally recovers immediately by an amount equivalent to γ_1, but the dashpot retains its final strain.

For a continuously varying shear stress, considering the elastic (e) and viscous (v) components separately:

$$\tau = G\,\gamma_e \qquad : \qquad \frac{d\gamma_e}{dt} = \frac{1}{G}\frac{d\tau}{dt} \qquad (3.3)$$

$$\tau = \eta\,\frac{dV_x}{dx} = \eta\,\frac{d\gamma_v}{dt} \quad : \qquad \frac{d\gamma_v}{dt} = \frac{1}{\eta}\,\tau \qquad (3.4)$$

$$\frac{d\gamma}{dt} = \frac{1}{G}\frac{d\tau}{dt} + \frac{1}{\eta}\,\tau \qquad (3.5)$$

If such a material is strained to a certain level γ^* and then held at this extension, $\dot\gamma$ becomes zero, so:-

$$\frac{1}{G}\frac{d\tau}{dt} + \frac{1}{\eta}\,\tau = 0 \qquad (3.6)$$

and the stress will relax according to the solution of Equation 3.6, viz.:-

$$\tau = \tau^*\exp(-Gt/\eta) = \tau^*\exp(-t/T) \qquad (3.7)$$

T can thus be considered as a relaxation time, being in fact the time required for the stress to decay to $1/e$ of its original value. The relaxation time is an important parameter for polymers. When $t \ll T$ the material is essentially elastic, but when $t \gg T$ it behaves like a liquid. (There is a close analogy with creep in metals at high temperatures.)

More complex polymers exhibit *relaxation spectra*, behaving differently at different strains as well as strain-rates or frequencies. These spectra are especially important for highly complex natural polymers such as skin and ligaments[1].

(b) The second simple model assumes the elementary spring and dashpot in parallel. The forces are now additive, and since there is in reality only one sample, the area is not relevant and the stresses can be added:

$$\tau = \tau_e + \tau_v = G\gamma + \eta\frac{\partial\gamma}{\partial t} \qquad (3.8)$$

$$\frac{d\gamma}{dt} + \frac{G}{\eta}\gamma = \frac{\tau}{\eta} \qquad (3.9)$$

This is solved by multiplying throughout by $\eta \exp(Gt/\eta)$, as follows:-

$$\tau \exp(Gt/\eta) = \gamma.G \exp(Gt/\eta) + \eta \exp(GT/\eta).\frac{d\gamma}{dt} =$$

$$\frac{d}{dt}\{\eta\gamma \exp(Gt/\eta)\} \qquad (3.10)$$

$$\tau \exp(Gt/\eta) \, dt = \eta\gamma \exp(Gt/\eta) \qquad (3.11)$$

$$(\tau\eta/G) \exp(Gt/\eta)+B = \eta\gamma \exp(Gt/\eta); \qquad (3.12)$$

Thus, at $\tau = 0$, $\gamma = 0$:-

$$\gamma = \frac{\tau}{G} \{1 - \exp(-t/T)\} \qquad (3.13)$$

A suddenly applied stress causes an exponential increase in the strain.   (The model beam cannot of course tilt, both components being contained in the same solid.)   If the stress is removed τ becomes equal to zero and the strain decays exponentially (Fig. 3.4), according to the relationship:-

$$\gamma = \gamma_0 \exp(-t/T) \qquad (3.13a)$$

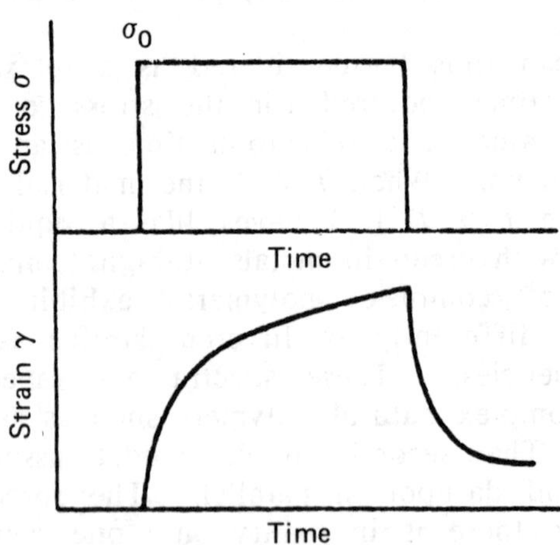

Fig. 3.4 The response of a Voigt model

If the component is held at a fixed strain γ_2, τ remains finite, in contrast to the relaxation of stress in the series model.

It is possible, of course, to combine the two different models and so to simulate more closely the behaviour of more complex polymers.

For example Fig. 3.5 shows a summation:-

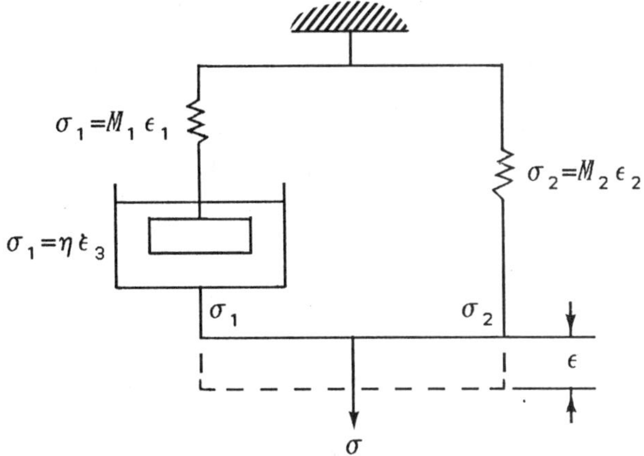

Fig. 3.5 A composite viscoelastic model

$$\epsilon = \frac{\sigma}{M_1 + M_2} \left\{ 1 + \frac{M_1}{M_2} \left[1 - \exp\left(-\frac{M_1 \cdot M_2}{M_1 + M_2} \cdot \frac{1}{\eta_0} \cdot t \right) \right] \right\} \tag{3.14}$$

3.3.2 *Temperature and Structure Dependence*

The intermolecular forces in non-polar polymers are predominantly weak van der Waals interactions, but many polymers are polar and exhibit much greater strength[5].

High-polymer structures may be amorphous, crystalline or orientated crystalline, as sketched in Fig. 3.6.

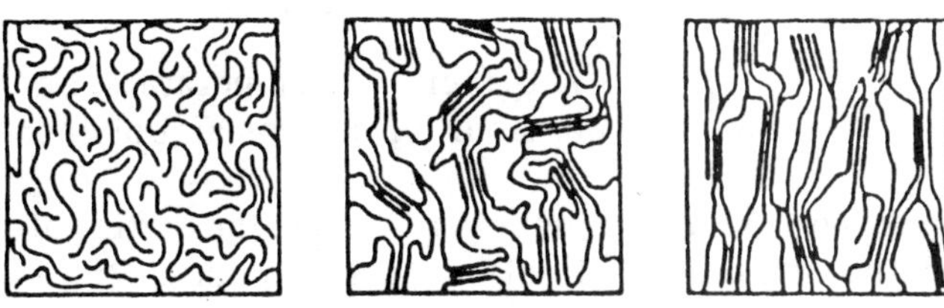

(a) amorphous, (b) crystalline, (c) orientated

Fig. 3.6 Schematic representation of
high-polymer macro-structures

The crystalline regions are usually isolated zones in which the molecules are locally aligned into an orderly pattern. This increases the density of packing and also the strength of interactions.   There is therefore a strong dependence of the strength of a polymer on its density, both being dominated by the degree of crystallinity.   The improved properties of HDPE have already been mentioned.

The crystalline regions themselves will become increasingly aligned as the polymer is stretched.   Nylon is a good example of the high strength that can be obtained in this way.

Increasing temperature tends as usual to increase randomization, so the proportion of amorphous material increases with temperature.   There is not a single-temperature melting point or liquidus, as with metals, at which crystals are in equilibrium only with a liquid phase.   Even moderate increases of temperature drastically reduce the strength of most polymers, although increasing the ductility, as shown in Fig. 3.7.

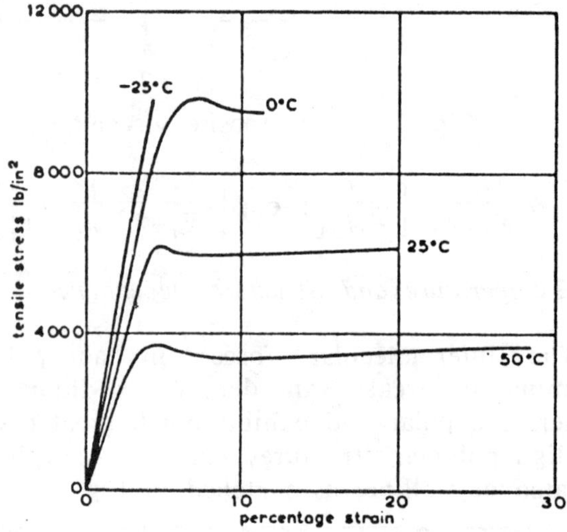

Fig. 3.7 Effects of temperature on the
behaviour of a typical polymer

The Young's modulus, being the ratio of elastic stress to elastic strain, will also depend upon the crystallinity and density.   It is an important descriptive parameter for polymers. The numerical values of E will vary considerably with chain length, crystallinity and temperature, unlike the relatively stable value for metals, which is also very much higher. Table 3.1 lists some representative values.   As will be seen later (in Section 3.9), the modulus of elasticity is a very important factor in reinforced composites.

TABLE 3.1 *Strength and modulus values*

Material	Impact J/m	Tensile N/mm²	Modulus kN/mm²	Specific gravity
Polyacetal	69.4	70	3.6	1.42
Polypropylene	21.4	35	1.2	0.92
Nylon 6/6	80.1	80	3.0	1.14
Polycarbonate	854.1	60	2.4	1.2
Silicone resin	2.14	34	-	1.9
Polyethylene	-	-	-	0.9
Polystyrene	-	-	-	-
Synthetic rubbers	-	-	-	-
GRP	117.4	121	8.4	1.52

3.3.3 *The Influence of Moisture*

Many polymers absorb moisture from the atmosphere, which significantly affects the intermolecular forces and hence the strength of the material, as shown in Fig. 3.8 for polyethylene.

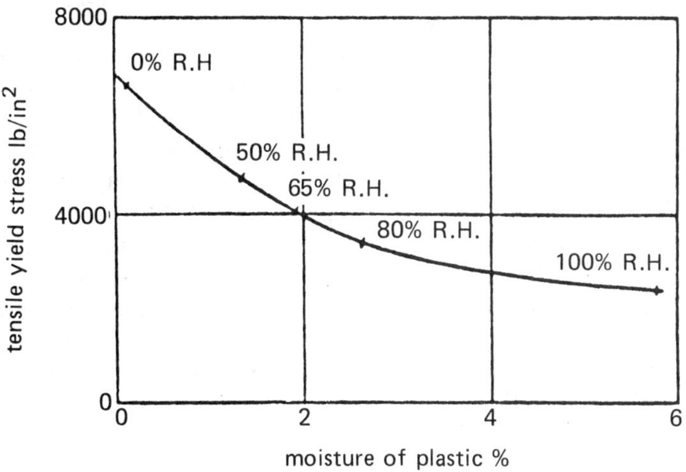

Fig. 3.8 The influence of humidity on strength

On a hot humid day (90% R.H.) the moisture content of air is about 4% by weight, so the yield strength is almost halved.

3.4 Silicone Polymers

The silicones are based on a backbone of Si and O atoms.

$$
\begin{array}{ccc}
R & & R \\
| & & | \\
-\,O\,-\,Si\,-\,O\,-\,Si\,-\,O\,- \\
| & & | \\
R & & R
\end{array}
$$

Here R represents an organic radical.   For example the mer of polydimethyl silicone is

$$
\begin{array}{c}
H \\
| \\
H\,-\,C\,-\,H \\
| \\
-\,Si\,-\,-\,O \\
| \\
H\,-\,C\,-\,H \\
| \\
H
\end{array}
$$

These materials are thus more correctly described as organosilicones.   They were developed to produce unreactive polymers that remain flexible over a much wider temperature range than the organics.   Silicone fluids remain fluid at very low, sub-zero, temperatures but boil at higher temperatures than hydrocarbon oils.   Silicone rubbers are elastic even at $-80\,°C$ but can be used up to $500\,°C$ and may be resistant to brief exposure to a flame at $2800\,°C$.

As with organic polymers, the chain length may be very large, e.g. 10,000 SiO repeats.   Cross-linking and the introduction of side chains can produce greater rigidity and strength.   The viscoelastic properties of silicone rubber ("bouncing putty") give a vivid demonstration of the influence of strain rate.

3.5 Filled Polymers

Polymers are often used with various powdered or fibrous fillers some of which simply economize on the quantity of resin required, but these *diluent fillers* are less important than the *reinforcing fillers* that improve the mechanical properties.

Common fillers are wood flour, cellulose, cotton flock,

paper or linen. Mica or asbestos may be used to improve heat resistance.

A quite different function is performed by plasticizers. These are low molecular weight organic liquids of high boiling point, added to separate the polymer molecules and to facilitate deformation.

Fibre reinforced polymers such as GRP (glass reinforced polymer) will be discussed in the context of composites (see Section 3.9).

3.6 Elastomers and Rubbers

3.6.1 *Natural Rubber*

Natural rubber is obtained from the tree *Hevea brasiliensis*. It contains the cis-isomer of polyisoprene, which is a kinked molecule, and the rubber remains amorphous unless severely strained. Gutta percha, in contrast, contains the smoother trans-isomer and is a highly crystalline polymer[6]. Rubber is weak at normal temperatures and deforms easily by molecular sliding. It can, however, be substantially strengthened by cross-linking and other means. The most common method is vulcanizing, discovered by Goodyear in 1839, which involves heating with sulphur to break the double bonds and introduce sulphur cross-links.

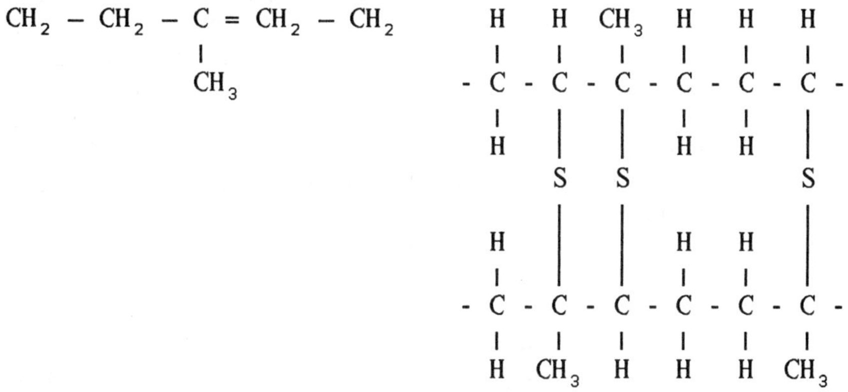

$$CH_2 - CH_2 - C = CH_2 - CH_2$$
$$|$$
$$CH_3$$

Fig. 3.9 Vulcanization of rubber (polyisoprene)

Automobile tyre treads may contain up to 5% S but a hard rubber battery case may contain as much as 40%. The tensile strength can be increased tenfold (up to 20 MPa) and the useful temperature range extended between -40C° and

+100 °C by vulcanizing.

3.6.2 *Synthetic Rubbers*

Many synthetic rubbers have been produced, usually as by-products of coal and petroleum[7]. These are mainly co-polymers; for example, Buna N is formed from butadiene and acrylonitrile. It has very good resistance to abrasion and to ageing, and is unaffected by petroleum. Neoprene is a chlorobutadiene polymer, which is much more resistant to sunlight and oil than natural rubber. Fluorocarbon rubbers have excellent high temperature properties but they are very expensive. Viton A is a co-polymer of perfluoropropene and vinylidene fluoride which is chemically inert and is stable up to 300 °C. Silicone rubbers combine the high temperature stability of silicones with the solvent resistance of fluorocarbons.

3.6.3 *Elastomers*

Elastomers have many of the properties of rubbers but have quite different chemical composition. They include polyethylenes, polyesters, polyvinyls and some silicones, sometimes halogenated to provide resistance to hydrocarbons[8].

It is important to recognize that the elastic behaviour of these materials is not the simple Hookean linear type with strain directly proportional to stress. The apparent modulus will increase with strain, and the Poisson's ratio may rise towards 0.5. This variation in elastic modulus is still more pronounced in protein rubbers such as resilin, which is a remarkable natural material occurring in the wing system of insects. There is much to be learnt from the effective use of variable modulus, variable hysteresis elastomers and polymers in nature[9].

3.6.4 *Transitions in Polymers*

It is convenient to distinguish elastomers, rubbers, soft polymers and glassy organic compounds, but these really form a continuum of viscoelastic solids of various complexities (Fig. 3.10). A single polymer, such as polystyrene may exhibit all these properties in turn as the temperature is varied.

Below the glass transition temperature the materials are hard and stiff, with Young's modulus values of around 100 MPa. As the temperature is raised, a material becomes leathery and then rubbery until near its breakdown temperature when it behaves like a viscous liquid.

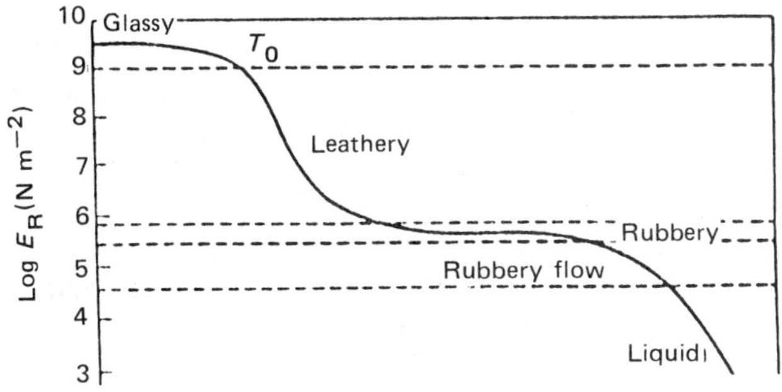

Fig. 3.10   The effect of temperature on relaxation modulus for polystyrene[1,10]

Some typical physical and mechanical properties of a selection of important polymers, elastomers and rubbers are set out in Table 3.2.

TABLE 3.2 *Physical properties of some important polymers*

Material	Specific gravity	Thermal conductivity $\frac{cal\text{-}cm}{°C.cm^2.sec}$ at 68°F	Thermal expansion in/in/°F at 68°F	Electrical resistivity, ohm-cm at 68°F	Average modulus of elasticity,psi at 68°F
Melamine-formaldehyde	1.5	0.0007	15 X 10^{-6}	10^{13}	1.3 X 10^6
Phenol-formaldehyde	1.3	0.0004	40 X 10^{-6}	10^{12}	0.5 X 10^6
Urea-formaldehyde	1.5	0.0007	15 X 10^{-6}	10^{12}	1.5 X 10^6
Rubbers (synthetic)	1.5	0.0003	–	–	500-10,000
Rubber (vulcanized)	1.2	0.0003	45 X 10^{-6}	10^{14}	0.5 X 10^6
Polyethylene	0.9	0.0008	100 X 10^{-6}	10^{13}	–
Polystyrene	1.05	0.0002	35 X 10^{-6}	10^{18}	0.4 X 10^6
Polyvinylidene chloride	1.7	0.0003	105 X 10^{-6}	10^{13}	0.05 X 10^6
Polytetrafluoroethylene	2.2	0.0005	55 X 10^{-6}	10^{16}	–
Polymethyl methacrylate	1.2	0.0005	50 X 10^{-6}	10^{16}	0.5 X 10^6
Nylon	1.15	0.0006	55 X 10^{-6}	10^{14}	0.4 X 10^6

3.7 Glasses

Inorganic glasses are familiar domestic materials, based on silica with the addition of sodium, calcium, potassium, boron, or other oxides.

Glasses have no well-defined melting point and can correctly be described as supercooled liquids. The instability usually associated with supercooling is not normally observable, but as is well known, ancient glasses *devitrify* or *crystallize*.

The most common window and bottle glass is soda-lime, made from silica sand with the addition of NaO and CaO. Lead glasses, also known as flint glass, contain PbO and are valued for their high refractive index. Laboratory glassware is usually borosilicate, containing SiO_2 and B_2O_3. Because this has a low coefficient of thermal expansion it is more resistant to thermal shock. Pyrex and related glasses are used for ovenproof kitchen ware.

Glass is obviously valuable for its transparency but one of the main uses in manufacturing processes is, surprisingly, as a lubricant. The softening of silicate glass to a highly viscous liquid provides excellent hydrodynamic conditions for lubrication at temperatures of 1000°C-1200°C, especially for the extrusion of stainless steel or nickel alloys[11].

Glass fibre is widely used as a reinforcing material for polymers. It has a very high modulus compared with polymers and in a freshly drawn condition it exhibits high strength[12].

3.8 Refractory Materials

3.8.1 Applications of Materials at High Temperatures

Throughout history there has been a link between advances in civilization and the availability of increasingly higher temperatures for producing and working metals and alloys[13]. During the second half of the twentieth century the demands for high-efficiency energy-conversion have made it imperative to provide greater strength and toughness at ever-increasing temperatures. Table 3.3 shows some applications and characteristic conditions.

TABLE 3.3 *Approximate maximum temperatures (°C)*
required by various industrial users

Power and propulsion	Mineral and chemical processing	Energy dissipation
Steam turbine 650	Petroleum 1000	Rocket chamber 650
Gas turbine 1400	Silicon carbide 2200	Re-entry vehicle 16000
Chemical rocket 4000	Graphite 3000	
Gas-cooled reactor 16000	Ceramics 3500	

There are no theoretical limits to the temperatures that can be generated. All chemical compounds dissociate by about 6000 K but total ionization of all elements may require 10^7 K and the thermal energy required to exceed nuclear bond energies is estimated to be equivalent to about 10^{20} K.

In practice however the important upper limit for structural materials 'is set by melting. Hafnium carbide is the most refractory known compound, melting at 4160 K. Tungsten melts at 3663 K, iron at 1803 K and aluminium at 933 K[14].

There is a general rule, related to the self-diffusion energy, stating that materials cannot be usefully employed for long periods at temperatures in excess of about 2/3 of their melting point (T_m K). This is a useful guide for many purposes, but it should be recognized that sometimes only a very short operating life is required, for example in rockets and re-entry vehicles. In these special circumstances, polymers can be used even when the temperatures at hypersonic re-entry may reach 8000 K. By sacrificial ablation, large amounts of energy can be dissipated rapidly.

Most high temperature materials are metals, alloys or ceramics. The latter word is derived from the Greek "$\kappa\eta\rho\alpha\mu o\sigma$", (meaning potter's earth) but for engineering purposes it is more clearly defined in terms of oxides, borides and nitrides, hardened by firing. Sometimes some carbides, silicides and sulphides are included in this category.

3.8.2 *Categories of Refractory Materials*

3.8.2.1 *Refractory bricks* have for centuries been used in furnaces where strength is not a major consideration. Fire clays consist of silica and alumina with small proportions of iron oxide, lime and magnesia. These will resist temperatures in excess of 1200 °C, and in general, higher alumina content is associated with higher softening point. Silica bricks containing up to 95% SiO_2, and natural and synthetic graphite bricks are also used[2]. All these materials are much weaker than the more recently developed engineering ceramics described below.

3.8.2.2 *Heat-resistant steels,* usually having a low alloy content of Cr, Mo and V, often with small additions of B or Nb, can be used at temperatures up to 1200 °C. They have stable structures and are therefore suitable for use in industrial plant where service lives of many years may be required. Higher alloys of 18Cr/8Ni type, stabilized with Ti, exhibit stainless properties and are used where corrosion resistance is also required[15,16].

3.8.2.3 *Superalloys* are mainly used for aircraft turbines, where they now account for some 70% of the total engine weight. In the early gas turbines the maximum temperatures were about 820 °C, and Ni or Co based alloys of the Hasteloy type were used.   These alloys have been considerably improved since 1940 but as they rely on precipitation hardening they are fundamentally unstable in structure and must not be overheated.   A measure of the improvements achieved is given by the increase in thrust/weight ratio by a factor 3 in the modern turbines, together with improved fuel consumption and an increase in average overhaul time from 100 hrs to 12000 hrs.   The Nimonic range of alloys in particular has also found wide applications in industry[17].

Fig. 3.11 indicates dramatically the progress that has been made in the development of superalloys[18].

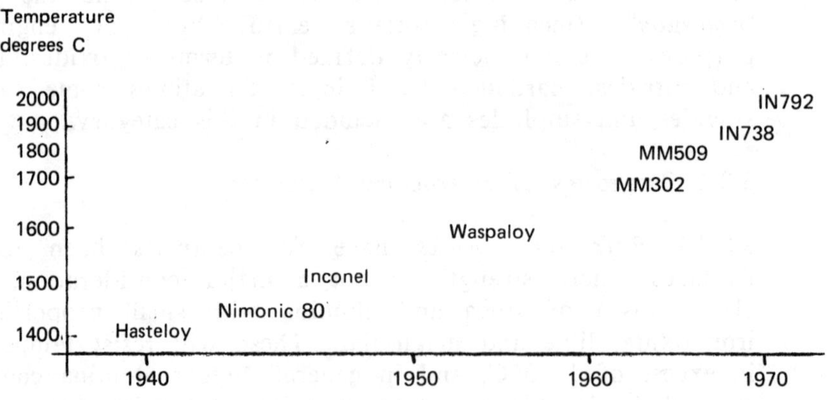

Fig. 3.11 The temperatures for 100 hrs survival under a tensile stress of 140 MPa (after SIMS and HAGEL[18])

It may be possible to increase the strength of current alloys by a factor of as much as 2 in the temperature range of 450 °C–750 °C by further attention to grain size refinement and to directional solidification.

3.8.2.4 *Refractory metals*

Refractory metals and their alloys have potential uses at high temperatures, though they tend to be brittle at lower temperatures. Tungsten can be used unalloyed, up to 1650°C. Molybdenum alloys such as TZM (0.5Ti, 0.08Zr, 0.02-0.08C) will also withstand 1600°C, in the absence of an oxidizing atmosphere. Niobium alloys are somewhat less refractory, but may increase greatly in importance because the element is more abundant. Tantalum, in contrast, is very scarce, but can in principle be used at temperatures up to 2900°C.

Although not a metal, graphite should be included for consideration. It has an excellent strength/weight ratio over a very wide temperature range. Uniquely, the strength of graphite increases substantially as the temperature is raised[19].

3.8.2.5 *Oxide ceramics*

Oxide ceramics are the compounds most commonly implied by the term ceramics in an engineering context. Alumina is very widely used, in pure form as α-Al_2O_3 with a rhombohedral structure, or alloyed with Cr_2O_3 for cutting tools[20]. A cubic form (γ-alumina) also exists and is used, for example, as a polishing abrasive.

TABLE 3.4 *Mechanical properties of high-alumina ceramics*

	79%Al_2O_3	88%Al_2O_3	95%Al_2O_3	100%Al_2O_3
Density kg/m³	2962	3432	3515	3764
Coefficient of linear expansion × 10⁶ per°C (to 100°)	4.4	5.33	3.6	6.6
Softening temperature °C	1649	1971	1927	2055
Compressive strength MPa	830.8	995.6	1290	2896
Transverse rupture strength MPa	160.7	201.3	248.7	275.8-517.1

The strength of alumina depends greatly upon its purity as shown in Table 3.4. It is apparent that the properties improve considerably even between 95 and 100 per cent, but this has to be balanced against the high cost of purification. In the extreme, single crystals of pure alumina are prepared as transparent synthetic sapphire. This can be ground to shape and size with very high precision, even to make ball bearings. A second factor exerting a dominant influence on the properties of alumina and other ceramics is the porosity, often expressed as the percentage departure from theoretical density. Fig. 3.12 shows the effect on the life of ceramic cutting tools. It should be noted that this graph covers only the range from 99 to 99.99% of the theoretical density. The measured permeability may also be very low, since the voids produced in the manufacture will not usually be connected.

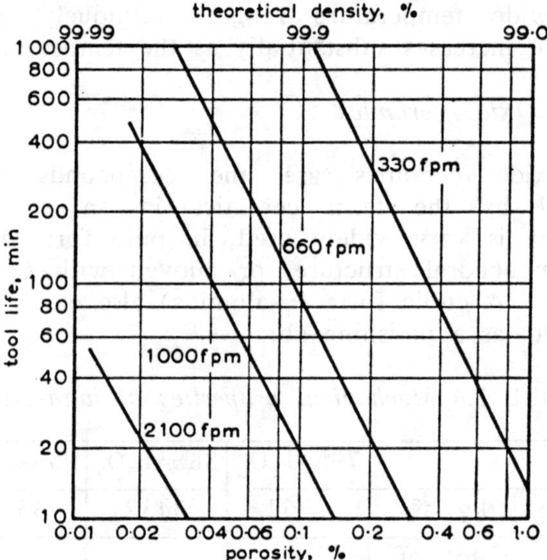

Fig. 3.12 The effect of porosity on tool life
(after BREWER[20], by courtesy of *Engineer's Digest*)

It is commonly believed that the strength and especially the toughness of alumina will be reduced if large grains are permitted to grow, though this conclusion has been challenged[21,22].

Zirconia (ZrO_2) is another industrially very important oxide ceramic. In its pure form, it has an f.c.c. structure that transforms to tetragonal at high enough temperatures. This involves a volume change and is therefore highly undesirable. The cubic structure can be stabilized by the addition of CeO and Y_2O_3. This allows the oxide to be used at temperatures up to 2400°C. Zirconia exhibits higher strength than alumina at 1100°C, as shown in Table 3.5. The low thermal expansion is of particular importance under conditions where thermal shock may be involved, as will be discussed later.

TABLE 3.5 *A comparison between stabilized zirconia and alumina*

Oxide	Melting point °C	Bond strength 25°C	1100°C	Compressive strength 25°C	1100°C	Thermal expansion at 1200°C (%)
Al_2O_3	2000	280	120	2400	600	1.2
ZrO_2	2770			2400	800	0.7

Thoria (ThO_2) and beryllia (BeO) have also been considered for high-temperature applications, for example for gas turbine blades. BeO has a very low coefficient of thermal expansion, which makes it attractive for this purpose, despite the toxicity of finely divided beryllia. Both have substantially lower compressive strength than alumina or zirconia, as indicated in Fig. 3.13.

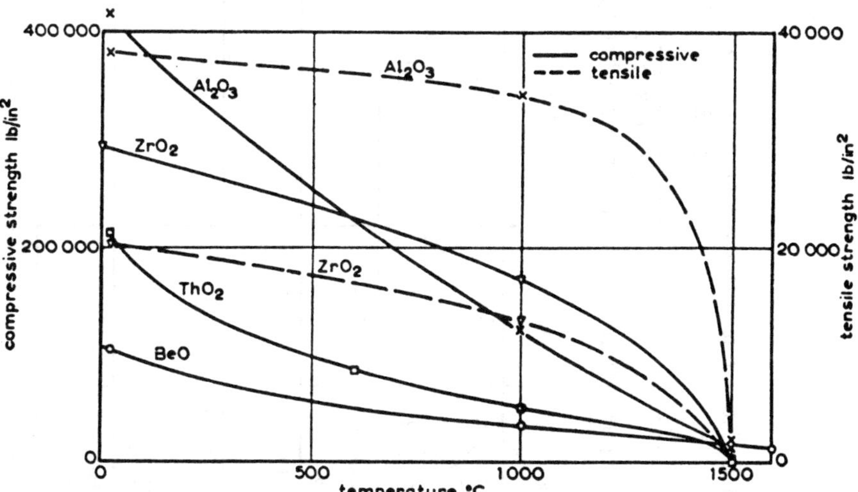

Fig. 3.13   Temperature dependence of tensile and compressive strength of four important oxides[23]

Silica (SiO_2) is not usually considered to be a refractory oxide, although it melts at 1726°C. It is widely used in its vitreous form, as a glass. Zircon (ZrO_2-SiO_2) melts at 2420°C and has a very low thermal expansion and correspondingly excellent resistance to thermal shock. The properties of various compound oxides have been summarized by BURDICK et al.[24] as shown in Table 3.6.

TABLE 3.6 *The mechanical properties of some oxide materials (after BURDICK et al.[24])*

Designation of material	Composition mole	Density ρ g/cm^3	100-h stress-rupture strength at 980°C σ_{980} N/mm^2	Approx ratio σ_{980}/ρ	Stress for min^m: creep rate of 0.0001 % per h. at 870°C σ_{870} N/mm^2	Approx: ratio $\dfrac{\sigma_{870}}{\rho}$
4811 C	48BeO-2Al$_2$O$_3$-ZrO$_2$ (+2%CaO)	3.0	117	39	97	32
151	Mgo-5BeO-ZrO$_2$	3.8	117	31	--	--
353	3MgO-5BeO-3ZrO$_2$	4.4	124	28	--	--
358	3MgO-5BeO-8ZrO$_2$	4.9	117	24	--	--
16021 T	160BeO-2Al$_2$O$_3$-ThO$_2$ (+2%TiO$_2$)	3.0	41	14	90	30
163	MgO-6BeO-3ZrO$_2$	4.4	110	25	--	--
Sillimanite	3Al$_2$O$_3$-2SiO$_2$	2.8	38	14	--	--
Superalloy	Proprietary	8.31	70	8.4	80	9.6

3.8.2.6 *Carbides, borides, nitrides and oxy-nitrides*

The *carbides* are in general very hard, as can be seen from Table 3.7, and indeed are often known as hardmetals (from the German *Hartmetall*). They have been very widely used for the tools in cutting and forming processes, especially those made from the bonded forms of WC-Co and WC-TiC-Co. They are not very resistant to high temperature oxidation or corrosion, but the fused or sintered carbides can be used in some circumstances.

TABLE 3.7   *Hardness and strength of some carbides*

Carbide	Microhardness (kg/mm^2)		Transverse rupture strength (MPa)	
	20°C	900°C	1400°C	2000°C
WC	1900	1300		
Cr C	1600	700		
Ti C	3500	400		
B C			500	200
Si C			450	

Boron carbide shows very good strength at the highest temperatures. SiC, alone amongst the carbides, can be used in air because it forms a coherent protective silica (SiO_2) coating when heated in the presence of oxygen. It is used for heaters, nozzles and other similar structures.

Although diamond is not usually regarded as a refractory because it will burn in air, it is the hardest known material and can be very effectively used for drawing-dies and cutting-tools. Recently, excellent performance has been obtained from polycrystalline diamond (PCD) tools[25].

Most *borides* are hard and stable. They oxidize only above about 1300°C, and they have low volatility so they can be used *in vacuo* at 2500°C. Ti, Zr, Hf and Ce borides are valuable for crucibles, and rocket nozzles have been made from ZrB.

Nitrides have come into prominence recently as structural materials, the hardness and strength of some being shown in Table 3.8. Naturally occurring BN has a hexagonal structure, similar to that of graphite. It is very weak and can be used as a high temperature lubricant but the artificial cubic structure, analogous to that of diamond and customarily referred to as CBN, provides high strength and hardness.

TABLE 3.8   *Hardness and strength of some nitrides*

Nitride	Microhardness at 20°C (kg/mm^2)	Transverse rupture strength (MPa)	
		1200°C	1800°C
CBN			600
Si_3N_4	1800	200	
Sialon	1800	800	

CBN is used as a very hard abrasive. Silicon nitride in reaction bonded form (RBSN) is oxidation resistant, because of the protective SiO_2 formed, and is uniquely resistant to thermal shock because of its extremely low thermal expansion. Si_3N_4 can also be used to bond silicon carbide.

An important advance has been made recently in producing the quaternary compound Sialon, a silicon aluminium oxy-nitride[26]. This compound improves on the hardness and strength of silicon nitride and offers a new family of high temperature materials with outstanding thermal shock resistance. It retains good mechanical properties to higher temperatures than RBSN. A noteworthy feature is that all four elements are found in abundance all over the world.

3.8.2.7 *Silicides and sulphides*

Silicides are also stable at high temperatures in air, because of the protective coherent silica film. These compounds are not found in nature. $MoSi_2$ has a high electrical resistance and is used for furnace elements at 1200°C or more, but it is susceptible to both thermal and mechanical shock.

Most *sulphides* have no useful structural properties, but many are refractory. Ce-Th and U sulphides are used for crucibles up to 1900°C. CeS wires have been suggested for use in casting refractory turbine blades that contain ventilating holes. They can easily be dissolved out of the holes, using acids.

3.8.2.8 *Intermetallic alloys and cermets*

These can have good high temperature properties, in conjunction with oxidation resistance. For example, NiAl has a transverse rupture strength of 350 MPa at 950°C. It can also show a valuable ductility, reaching 40% elongation in a tensile test at 900°C under a stress of 70 MPa. When appropriately combined, the cermets can utilize the strength of ceramics and the ductility of metals. TiC bonded with Ni was developed for use in gas turbines at 850°C although it was eventually displaced by the Nimonic alloys. $Cr-Al_2O_3$ and MgO-TiN-NiO are promising materials. Cermets are used for cutting tools and have proved useful in certain aircraft brakes[27].

3.8.3 *Behaviour of Refractory Materials in Service*

3.8.3.1 *Creep*

One of the most important properties of a refractory material is its resistance to creep. In tensile applications the tendency to flow under stress at high temperatures causes elongation, for example, of a close tolerance turbine blade, and it also reduces the cross section, thereby increasing the stress level. Most creep tests are conducted under constant load but the interpretations are usually given in terms of constant stress. Fig. 3.14 shows a typical high temperature creep curve.

Such curves generally have up to three distinct regions. In the primary or transient creep régime the strain rate diminishes until it reaches a constant value at which steady state creep begins. The latter régime is the important one in practice and the linear slope is the creep rate usually quoted. Finally, if the test is conducted at constant load, the specimen will become unstable and fracture will follow shortly thereafter, in the tertiary régime.

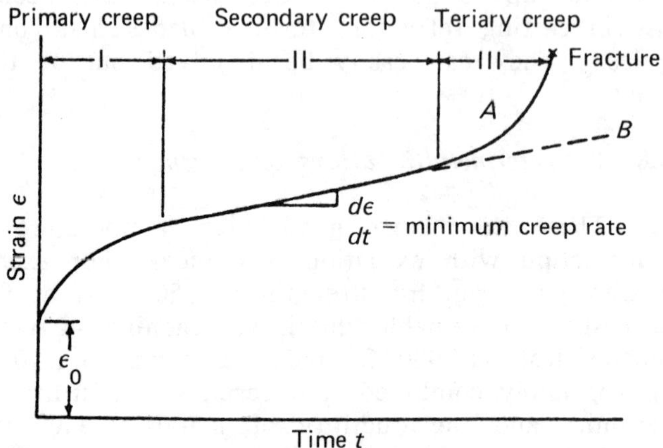

Fig. 3.14 Typical creep curve

The first two stages of creep are often represented by Equation 3.15, proposed originally by ANDRADE[28]:-

$$L = L_0(1+\beta t^{1/3}) \; \exp \; kt \qquad (3.15)$$

The behaviour is also sometimes described in terms of a spring and dashpot model, as used for the elastic viscous deformation of polymers.

Steady state high temperature creep can be interpreted as a balance between strain hardening which causes an *increase* in flow stress of amount $(\partial\sigma/\partial\epsilon)d\epsilon$, and thermal recovery which *reduces* the flow stress by an amount $(\partial\sigma/\partial t)dt$. In a test conducted under constant stress, $d\sigma=0$, and since the stress at constant temperature is a function of strain and time only, the following analysis applies:-

$$d\sigma = \left[\frac{\partial\sigma}{\partial\epsilon}\right]d\epsilon + \left[\frac{\partial\sigma}{\partial t}\right]dt = 0 \qquad (3.16)$$

$$\frac{d\epsilon}{dt} = -\left[\frac{\partial\sigma}{\partial t}\right]\Big/\left[\frac{\partial\sigma}{\partial\epsilon}\right] \; . \qquad (3.17)$$

which is a constant for steady state creep.

It is now generally accepted that creep is a thermally activated process, so the strain rate can be represented by an Arrhenius type of equation:-

$$\dot{\epsilon} = K \; \exp(-Q/RT) \qquad (3.18)$$

where K is a function of $(\partial\sigma/\partial\epsilon)$ and Q is an activation energy.

The time parameter t can conveniently be replaced by a *temperature-compensated time* parameter θ, such that:-

$$\theta = t \, \exp(-\Delta H/RT) \qquad (3.19)$$

where ΔH is the activation energy for the particular creep mechanism involved.

The strain can then be described in terms of two parameters only, the stress σ_c causing creep and the parameter θ, instead of having to describe the stress as a function of the three parameters ϵ, t and T. Thus:-

$$\epsilon = f(\theta,\sigma_c) \qquad (3.20)$$

and

$$\dot{\epsilon} = \frac{d\epsilon}{dt} = \left[\frac{\partial f}{\partial\theta}\right]\cdot\left[\frac{\partial\theta}{\partial t}\right] = f'\cdot\left[\frac{\partial\theta}{\partial t}\right] = f'\cdot\exp\,(-\Delta H/RT) \qquad (3.21)$$

where:-

$$f' = \frac{\partial f}{\partial\theta}.$$

The steady state creep rate is the minimum value, corresponding to θ_{min}, so:-

$$\dot{\epsilon}_{min} = f'(\theta_{min},\sigma_c)\,\exp(-\Delta H/RT) \qquad (3.22)$$

$$f'(\theta_{min},\sigma_c) = \dot{\epsilon}_{min}\,\exp(+\Delta H/RT) = Z \qquad (3.23)$$

defining a new parameter Z, known as the *Zener-Holloman* parameter[29].

Since θ_{min} is itself a function of σ_c only, K being a constant, the stress to cause creep can be written:-

$$\sigma_c = F\{\dot{\epsilon}_{min}\,\exp(+\Delta H/RT)\} = F\{Z\} \qquad (3.24)$$

The Zener-Holloman parameter, together with certain other related functions, form a good basis for the correlation of creep data, as pointed out by JONAS[30].

3.8.3.2 *Creep tests and stress-rupture tests*

A typical creep test is conducted by loading a tensile sample in a constant temperature enclosure and recording the extension over a period of 2,000 hours, or sometimes 10,000 hours. The calculated strain at nominally constant stress is plotted as shown in Fig. 3.15.

At high stress levels or very high temperatures, the steady-state creep rate increases and the range between primary and tertiary creep diminishes. The *creep strength* is defined for design purposes as the stress which will produce 1% strain in 10,000 hours for jet engines and related purposes, or 1% strain in 100,000 hours (11–12 years) for steam turbines.

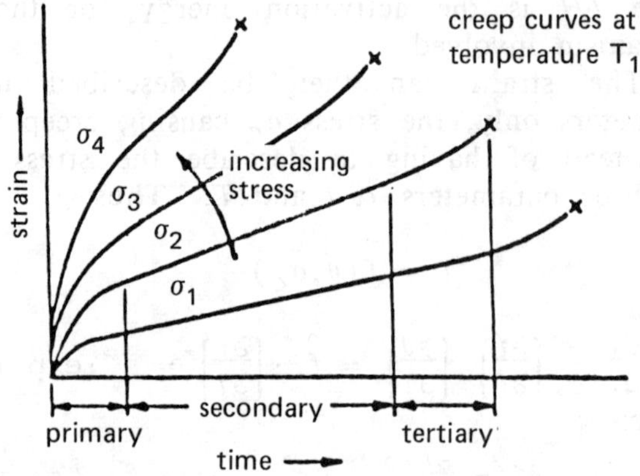

Fig. 3.15 Creep curves at various stress levels

These values are obtained from experimentally established families of graphs relating stress with the log creep rate at various temperatures, as illustrated in Fig. 3.16.

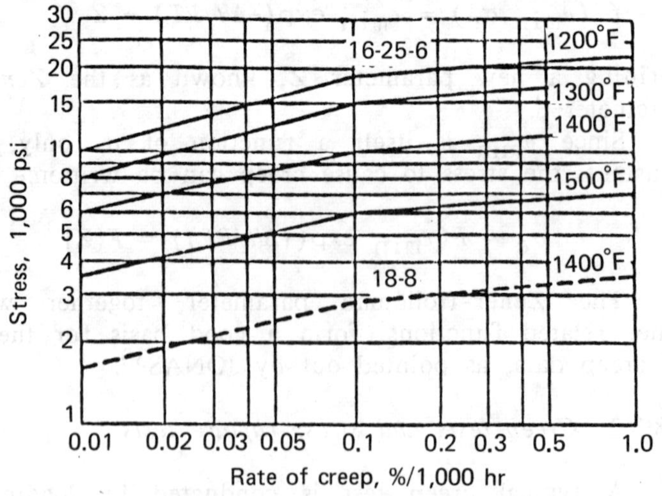

Fig 3.16 Stress vs. strain rate (%/1,000 hrs)
for various temperatures

These tests are necessarily slow and very expensive to carry out. To reduce the testing time, it is common practice to use a *stress-rupture test*. This is conducted at a higher stress level than a true creep test and may involve much larger strains, even as high as 50%. The time to produce failure at constant load and temperature is recorded, as shown in Fig. 3.17. The elongation is measured continuously so that the minimum creep rate can be determined. Such tests are widely used for jet engine alloys. The results can be correlated conveniently by using a *Larson-Miller*[31] parameter $T(c + \log t)\times10^{-3}$, or a related *Manson-Haford*[32] parameter $(T-T_a)/(\log t - \log t_a)$ in some instances.

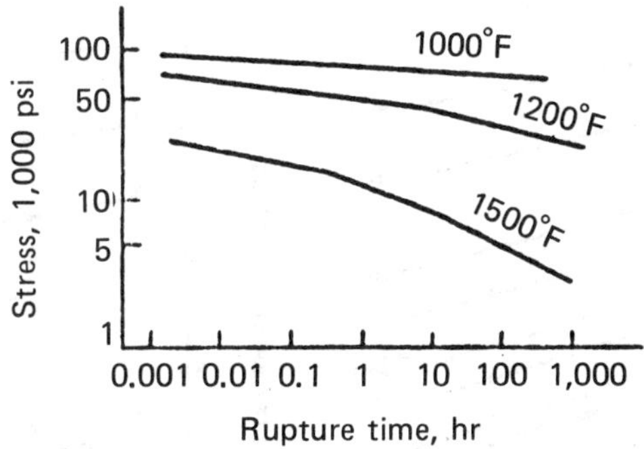

Fig. 3.17 Typical stress-rupture test results

It is found that plotting $\log t$ versus T for various (constant) stress levels produces a set of straight lines that converge on the point $(\log t_a)$. These parameters can be used, with caution, for prediction of long-term properties[33,34]. It is preferable, however, to use Weibull or other predictive statistical techniques.

3.8.3.3 *Hot-hardness tests*

Indentation tests are non-destructive and therefore economical of material. A large number of tests can be performed on one small sample.

TABOR[35] has shown that ball indentation hardness results at various temperatures and times of loading can be converted using an Arrhenius equation. There is indeed a close correlation with the tensile creep behaviour although, of course, there is no tensile plastic instability. The activation energy can be calculated from linear plots of indentation area against $1/T$. Where comparisons have been made, there is reasonably good agreement with the values deduced from self-diffusion studies.

Fig. 3.18 shows some detailed measurements made with two (Zn-Al and Al-Si) superplastic alloys. The method has not yet been applied to refractory materials.

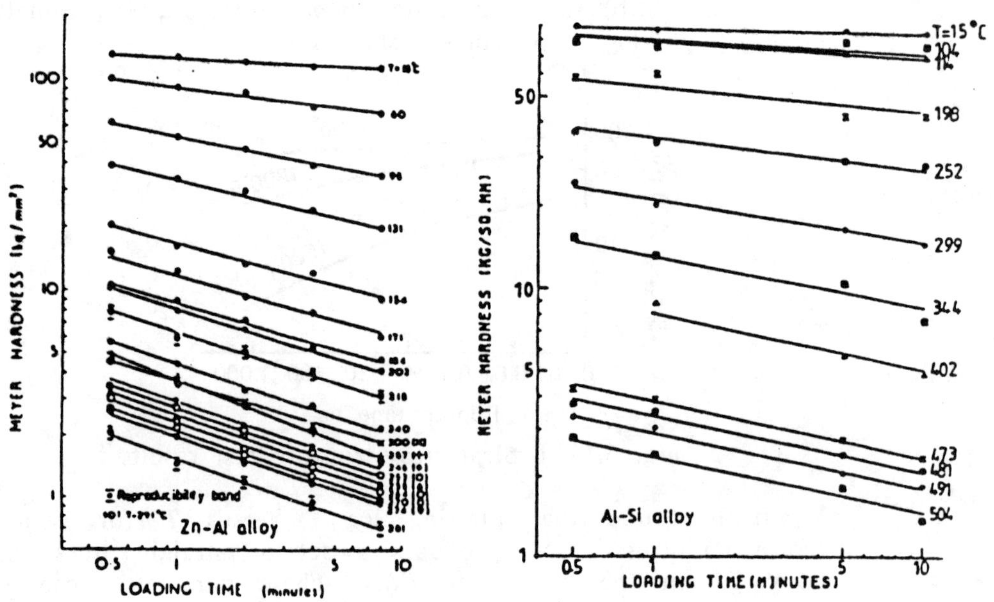

Fig. 3.18 Hardness/time temperature relationships
for two superplastic alloys[36]

3.8.3.4 *Stress relaxation*

In some instances, particularly for fixing bolts, creep becomes important in a different way. The dimensions of the parts being held together remain constant and the strain in the bolt is therefore constrained to stay unchanged, but the stress relaxes as creep occurs, leading to looseness of the joint.

3.8.3.5 *Fatigue*

At high temperatures the usual type of fatigue due to fluctuating mechanical stresses is of little significance because of the relaxation and damping due to creep. Thermal fatigue due to the stresses induced by unequal thermal expansion and contraction can, however, be a serious cause of failure.

Thermal strain is governed by differences in $\alpha \, \Delta T$ where α is the coefficient of thermal expansion and ΔT the temperature change. This is linearly related to the thermal stress σ_T by the Young's modulus E, e.g. in a bimetallic rod:-

$$\sigma_T = (\alpha_1 E_1 - \alpha_2 E_2) \, \Delta T \qquad (3.25)$$

The temperature difference between two adjacent zones is inversely related to the thermal conductivity K, so the thermal fatigue depends upon a parameter $(\sigma_f K)/(E\alpha)$, where σ_f is the fatigue strength at a mean temperature.

3.8.3.6 *Thermal shock*

In extreme circumstances such as quenching from a high temperature, the stresses may be high enough to cause failure even in one cycle, due to thermal shock. Resistance to this type of failure thus depends critically on thermal conductivity as well as on strength or toughness.

3.8.3.7 *Direct tensile failure. Toughness*

The response of a high temperature material to an applied load depends on both the load itself and the rate of loading. The stress can be expressed by a power law relationship to both strain and strain rate at given temperature as follows, for example:-

$$\sigma = k_1 \left[\epsilon^m \right]_{\dot{\epsilon}_1, T} + k_2 \left[\dot{\epsilon}^n \right]_{\epsilon_1, T} \qquad (3.26)$$

The values of both m and n vary strongly with temperature, as shown in Fig. 3.19. In recent years it has become widely accepted that stress alone is not a sufficient criterion for failure except in the simplest circumstances. Apart from the geometric effect of stress concentration, it is necessary also to consider the toughness of the material. As has been seen, the refractory materials range widely from ductile metals to brittle ceramics.

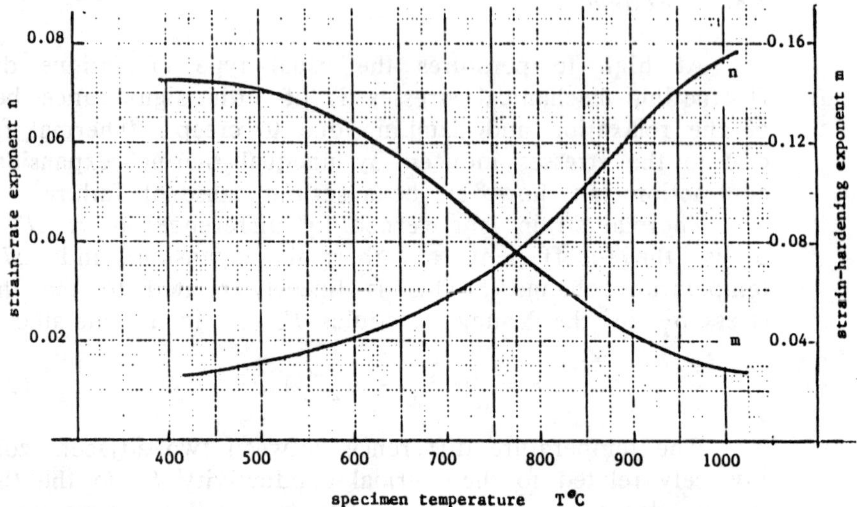

Fig. 3.19 Representative values of strain hardening exponent
m and strain-rate exponent *n* for chromium

Only the most recent data sheets include values for fracture toughness parameters such as K_{1C}, e.g. for Sialon, K_{1C} has the value 5 MNm$^{-3/2}$.

3.8.3.8 *Oxidation and corrosion*

All chemical reactions proceed more rapidly when the temperature is raised. This is important for most applications of refractories. Whenever air is present there is the possibility of oxidation, which will affect all but the oxide-protected materials or the oxides themselves. Many of the gases encountered may be corrosive and the effects of stress corrosion or environmental stress cracking should also be taken into account. Unfortunately, little is known about these at high temperatures.

3.9 Composites

It is convenient to deal with composites as a group because the principles are the same for reinforced concrete, fibreglass and resin, or the ferrite and cementite in steels, even for animal bones and tissues[1,37,38]. Possibly the earliest reference to the use of two materials to obtain better performance than from either separately relates to the making of clay bricks with straw[39].

The modern concept of a composite is that it is a new material dependent not only upon the nature of its constituents but also upon the way in which they are assembled[40]. This can be understood by considering first two simple models as illustrated in Fig. 3.20, in which sheets of two uniform materials are bonded firmly together and loaded respectively in series and in parallel.

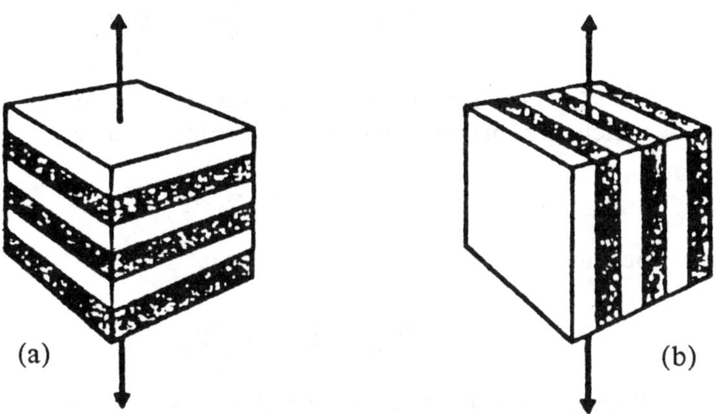

Fig. 3.20 The series (equal stress) and parallel
(equal strain) models of composites

3.9.1 *The Series Model* {Fig. 3.20(a)}

If the plates are loaded transversely to the interfacial planes, the force P will be transmitted equally through each. Since all have the same area A, the normal stresses σ_1 and σ_2 will also be equal and the following equations will apply:-

$$P = \sigma_c A; \qquad \sigma_1 = \sigma_2 = \sigma_c \qquad (3.27)$$

$$P = \sigma_1 A = \sigma_2 A \qquad (3.28)$$

The strains ϵ_1 and ϵ_2 will be related to the stresses by the respective Young's moduli E_1 and E_2, thus:-

$$\epsilon_c = \frac{\sigma_c}{E_c}; \quad \epsilon_1 = \frac{\sigma_1}{E_1} ; \quad \epsilon_2 = \frac{\sigma_2}{E_2} \qquad (3.29)$$

If $E_1 = E_2$ the strains are equal and there is no significant reinforcement, but if they differ the strain in the composite will be inversely proportional to the modulus of the composite.   It is therefore important to calculate this value from the individual moduli.

The total extension under load P is:-

$$\epsilon_c(L_1+L_2) = \epsilon_1 L_1 + \epsilon_2 L_2 \qquad (3.30)$$

where L_1 and L_2 are the total lengths of the two constituents, e.g. $L_1 = \sum l_1$. Substituting for the strains:-

$$\frac{\sigma_c}{E_c}(L_1+L_2) = \frac{\sigma_1}{E_1}L_1 + \frac{\sigma_2}{E_2} L_2 \qquad (3.31)$$

In practice it is convenient to consider volume fractions, which are proportional to the lengths, since:-

$$A_c = A_1 = A_2 \qquad (3.32)$$

hence:-

$$\frac{\sigma_c}{E_c}V_c = \frac{\sigma_1}{E_1}V_1 + \frac{\sigma_2}{E_2}V_2 \qquad (3.33)$$

Whence the modulus of the composite can be written as:-

$$\frac{1}{E_c} = \frac{1}{E_1}\frac{V_1}{V_c} + \frac{1}{E_2}\frac{V_2}{V_c} \qquad (3.34)$$

3.9.2 *The Parallel Model* {Fig. 3.20(b)}

If the bonded plates are loaded in the planes of the interfaces, the elongation of each component will be constrained to be equal to that of the other, and of course also to that of the composite. Thus:-

$$\delta L_1 = \delta L_2 = \delta L_c \qquad (3.35)$$

and since all are of equal length in the direction of the load application,

$$\epsilon_1 = \epsilon_2 = \epsilon_c \qquad (3.36)$$

and:-

$$\frac{\sigma_1}{E_1} = \frac{\sigma_2}{E_2} = \frac{\sigma_c}{E_c} = \epsilon_c \qquad (3.37)$$

In this model the stress in each component is directly proportional to its modulus. The respective loads carried are no longer equal:-

$$P_1 = A_1\sigma_1 = A_1\epsilon_1 E_1 ; \qquad P_2 = A_2\epsilon_2 E_2 \qquad (3.38)$$

$$P_c = A_c\epsilon_c E_c = P_1 + P_2$$

Thus

$$\frac{P_1}{P_c} = \frac{A_1 E_1}{A_c E_c} = \frac{A_1 E_1}{A_1 E_1 + A_2 E_2} ; \quad \frac{P_2}{P_c} = \frac{A_2 E_2}{A_1 E_1 + A_2 E_2} \qquad (3.39)$$

The modulus of the composite can be found from Equations 3.38:-

$$\epsilon_c A_c E_c = \epsilon_1 A_1 E_1 + \epsilon_2 A_2 E_2$$

$$E_c = \frac{A_1 E_1}{A_c} + \frac{A_2 E_2}{A_c} \qquad (3.40)$$

and for equal lengths this can be expressed in terms of the volumes:-

$$E_c = E_1\frac{V_1}{V_c} + E_2\frac{V_2}{V_c} \qquad (3.41)$$

which may be contrasted with Equation 3.34.

3.9.3 *General Reinforcement Model*

These simple models both show the importance of elastic modulus and can be extended to any number of components of different materials. The parallel and series models derived above give respectively upper and lower bounds to the modulus of the composite.

A general reinforcement can be envisaged as a set of fibres f, with some in parallel and the remainder in series, in a matrix m; or the actual orientations can be resolved into these two component directions. Suppose the volume fraction in parallel is x. Then $(1-x)$ will be in series, and from Equations 3.34 and 3.41:-

$$\frac{1}{E_c} = \frac{x}{E_f V_f/V_m + E_m V_m/V_c} + (1-x)\left[\frac{1}{E_f}\frac{V_f}{V_c} + \frac{1}{E_m}\frac{V_m}{V_c}\right] \qquad (3.42)$$

These formulations are valid only if both matrix and reinforcement remain elastic. If, for example, rubber is subjected to large strain, its modulus decreases. With metals the matrix stress may become large enough for plastic flow of the matrix to occur, and then the load-carrying capacity is limited since the effective plastic modulus $(d\sigma_m/d\epsilon)$ operates. This is much lower even than E_m so:-

$$E_c = E_f \frac{V_f}{V_c} + \frac{d\sigma_m}{d\epsilon} \frac{V_m}{V_c} \simeq E_f \frac{V_f}{V_c} \qquad (3.43)$$

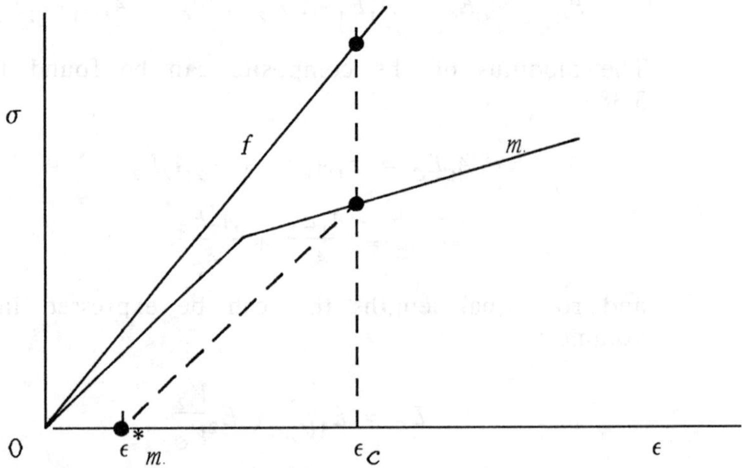

Fig. 3.21 A diagram illustrating the strains in an elastic fibre f and an elastic/plastic matrix m

On unloading this will result in residual stress, because the matrix cannot recover all the strain elastically. With reference to Fig. 3.21, when the composite is strained to some value ϵ_c, the fibres would recover completely on release of the load if unconstrained, whilst the elastic-plastic matrix would recover to a finite residual strain ϵ^*_m. Since the strains are in fact constrained to be equal, in the parallel model, an intermediate extension will result, with opposite residual stresses in the two components, leaving a tension in the high-modulus elastic fibres.

REFERENCES

(1) WAINWRIGHT, S.A., BIGGS, W.D., CURREY, J.D. and GOSLINE, J.M., *Mechanical Design in Organisms.* Edward Arnold, London, 1976.

(2) SEARLE, A.B. and GRIMSHAW, R. W., *The Chemistry and Physics of Clays.* Ernest Benn, London, 1959.

(3) WARD, I.M., *Mechanical Properties of Solid Polymers.* Wiley-Interscience, New York, 1971.

(4) BINGHAM, E. C., *Fluidity and Plasticity.* McGraw-Hill, New York, 1922.

(5) OGORKIEWICZ, R.M., *The Engineering Properties of Plastics.* Wiley-Interscience, 1977.

(6) VAN VLACK, L.H., *Elements of Materials Science.* Addison-Wesley, Reading, Mass:, 1959.

(7) SCHMIDT, A. and MARLIES, C. A., *Principles of High Polymer Theory and Practice.* McGraw-Hill, New York, 1948.

(8) SCHIEFER, H.M. and PAPE, P.G., "Uses of Silicones in Process Plants", *Chem. Eng. Aust.* 1982,7,49-52.

(9) GOW, B. S., and TAYLOR, M.G., "Viscoelastic Properties of Dog Arteries", *Air. Res.* 1968,23,111-122.

(10) AKLONIS, J.J., MACKNIGHT, W.J.L. and STEIN, M., *Introduction to Polymer Viscoelasticity.* Wiley, New York, 1972.

(11) SEJOURNET, J. and DELCROIX, J., "Glass Lubricant in the Extrusion of Steel", *Lubr. Engg.* 1955,11,389.

(12) GORDON, J.E., *The New Science of Strong Materials.* Pelican, London, 1982.

(13) TYLECOTE, R.F., *A History of Metallurgy.* Metals Society, London, 1976.

(14) CAMPBELL, J.E. and SHERWOOD, E.M., *High Temperature Materials and Technology.* Wiley, New York, 1967.

(15) LANSKAYA, K.L., *Heat Resistant and Refractory Steels.* Transtech, 1979.

(16) COLOMBIER, L. and HOCKMAN, J., *Stainless and Heat-resisting Steels.* Edward Arnold, London, 1967.

(17) BETTERIDGE, W. and HESLOP, J. (Eds.) *The Nimonic Alloys and other Ni-Base High-Temperature Alloys.* Edward Arnold, London, 1974.

(18) SIMS, C.T. and HAGEL, W.C., *The Superalloys.* Wiley, New York, 1972.

(19) ROWE, G.W., "High-Temperature Strength of Clean Graphite", *Nuclear Engg.* 1962,March,102-104.

(20) BREWER, R.C., "An Appraisal of Ceramic Cutting Tools", *Engineers Digest* 1957,18,381.

(21) ARCHIBALD, W.A. and SMITH, E.J.D., "Super-Refractories", *Brit. Ceram. Soc. Symp. "Ceramics", Stoke-on-Trent,* 1953,p536.

(22) CUTLER, I.B., "Strength Properties of Sintered Alumina in Relation to Porosity and Grain Size", *J. Amer. Ceram. Soc.* 1957,40,20.

(23) RYSHKEWITCH, E., *Oxide Ceramics.* Academic Press, New York, 1960.

(24) BURDICK, M.D., MORELAND, R.E. and GELLER, R.F., *Tech.Note 1561.* Nat. Adv. Cttee. Aeronautics, 1961.

(25) ASPINWALL, D.K., "Recent Developments in the Application of Poly-crystalline Diamond Tooling", *Proc.1st.Int.Mat. Removal Conf.,Detroit,* 1983,1-22. (S.M.E. MR 83-201.)

(26) ASPINWALL, D.K., TUNSTALL, M. and HAMMERTON, R., "Cutting Tool Life Comparisons", 25th *Int.M.T.D.R.Conf.,* 1985,269-276.

(27) HOVE, J.E. and RILEY, W.C. (Eds.) *Ceramics for Advanced Technologies.* John Wiley, 1965.

(28) ANDRADE, E.N.da C., "On the Viscous Flow of Metals and Allied Phenomena", *Proc.Roy.Soc.A.,* 1910,84,1.

(29) ZENER, C. and HOLLOMAN, J.H., "Effect of Strain Rate upon Plastic Flow of Steel", *J.App.Phys.* 1944,15,22-32.

(30) WONG, J.A. and JONAS, J.S., "Aluminium Extrusion as Thermally Activated Process", *Trans.Met.Soc.A.I.Min.E.* 1969,242,2271-2280.

(31) LARSON, F.R. and MILLER, J., "Time-Temperature Relationship for Rupture and Creep Stresses", *Trans.A.S.M.E.* 1952,74,765-775.

(32) MANJOINE, M.J., "Influence of Rate of Strain and Temperature on Yield Stresses of Mild Steel", *A.S.M.E. (J.App.Mech.)* 1944,11,211-218.

(33) HOSFORD, W.F. and CADDELL, R.M., *Metal Forming-Mechanics and Metallurgy.* Prentice-Hall, New Jersey, 1983.

(34) DIETER, G.E., *Mechanical Metallurgy.* McGraw-Hill, New York, 1961.

(35) TABOR, D., *The Hardness of Metals.* Oxford University Press, 1951.

(36) ROWE, G.W., "Mould Making in Superplastic Alloys", *Metals Society World* 1982,1(10),2-3.

(37) HOLISTER, G.S. and THOMAS, C., *Fibre-Reinforced Materials*. Elsevier, Amsterdam, 1966.

(38) CURREY, J.D., *Animal Skeletons*. Edward Arnold, London, 1970.

(39) THE BIBLE, Exodus, v.

(40) BROUTMAN, L.J., and KROCK, R.H. (Eds.) *Modern Composite Materials*. Addison-Wesley, 1967.

Chapter 4

THE INFLUENCE OF PROPERTIES
ON PRODUCTION PROCESSES

4.1 Introduction

The suitability of a particular material for a particular process is often difficult to assess. It usually depends on a special combination of properties that can only be brought into play by the process itself. Nevertheless, it is sometimes possible to single out certain over-riding characteristics that must be possessed by any material intended for a given process, or alternatively the specific properties of the material may determine the production route. For example a bar that is to be milled, turned, drilled or threaded should have good 'machinability', while a sheet that is to be drawn into a deep complex shell or dish should have good 'formability', or 'deep drawability'. Other materials, such as tungsten carbide for tool tips, can be produced only by sintering.

Unfortunately these descriptive labels are only loosely defined, largely in the context of process performance, and cannot easily be related to independent and closely controlled laboratory tests. In broad terms it is quite clear to a machinist that a leaded brass will have better machinability than a straight 60/40 brass. The lead is in fact added specifically for this purpose and is otherwise not a desirable feature. The chip will form well with low force, the surface of the workpiece will be good, and the tool will last well. Similarly, 'free-cutting' steel contains manganese sulphide for the same purpose but the detailed actions are not clearly understood and the only way to demonstrate the improved machinability is actually to cut under conditions resembling those to be used in practice. Even cutting at a substantially different speed can mask the desired characteristics. Again, deep drawability is a very important quality and special control is exercised in the production of such DD steel but it can be assessed only by actually drawing the sheet. In this instance laboratory-scale tests have been developed that correlate quite well with shop-floor performance, but they clearly simulate the actual process and require specialized equipment with close attention to such practical features as the lubrication and surface finish of the tools. Simple tensile elongation tests do not reveal the same features though there is a broad correlation with tensile 'ductility', itself as we have seen, a rather vague property which differs when measured as percentage elongation or as percentage reduction of area at the fracture.

Although the deficiencies of tests that simulate only part of the process are well known, many attempts have been made and are still being promoted with claims that this or that new simple test will assess machinability, formability, castability and so on.

In this chapter the characteristics of materials that appear to be important in particular processes will be described, and consideration given to some of the tests that can be used and the way in which the performance can be related to more fundamental properties. For convenience the subjects will be grouped as in the companion volume, *Manufacturing Technology, Vol. 2: Engineering Processes*[1].

4.2   Forming from the Liquid and Particle State

4.2.1   *Casting*

The most important property of a metal or alloy, with regard to forming from the liquid state, is its melting-point temperature or range of temperatures.

The equilibrium diagram of a representative binary system, copper-nickel, is shown in Chapter 2, Fig. 2.14. Referring back to that figure, the liquidus can be seen to be well defined as the temperature at which solidification starts as the alloy cools. With pure copper or pure nickel the metal becomes completely solid at this temperature, respectively at 1083° and 1452 °C, but for <u>any</u> alloy composition there is a range over which liquid and solid are in equilibrium together, at appropriate composition ratios. Complete solidification is achieved only at or below the solidus temperature.

In some specialized processes known as slush casting, rheocasting and squeeze forming, this intermediate temperature régime is utilized to obtain high density products by applying pressure during solidification[2], and it influences the crystal structure in conventional casting.

Copper and nickel are completely soluble in each other in

the solid state; but this is not typical. More commonly the alloy components are mutually soluble in the liquid state but insoluble or have limited solubility in the solid state[3]. The copper–silver system is representative of this group and shows a more complex diagram, Fig. 4.1.

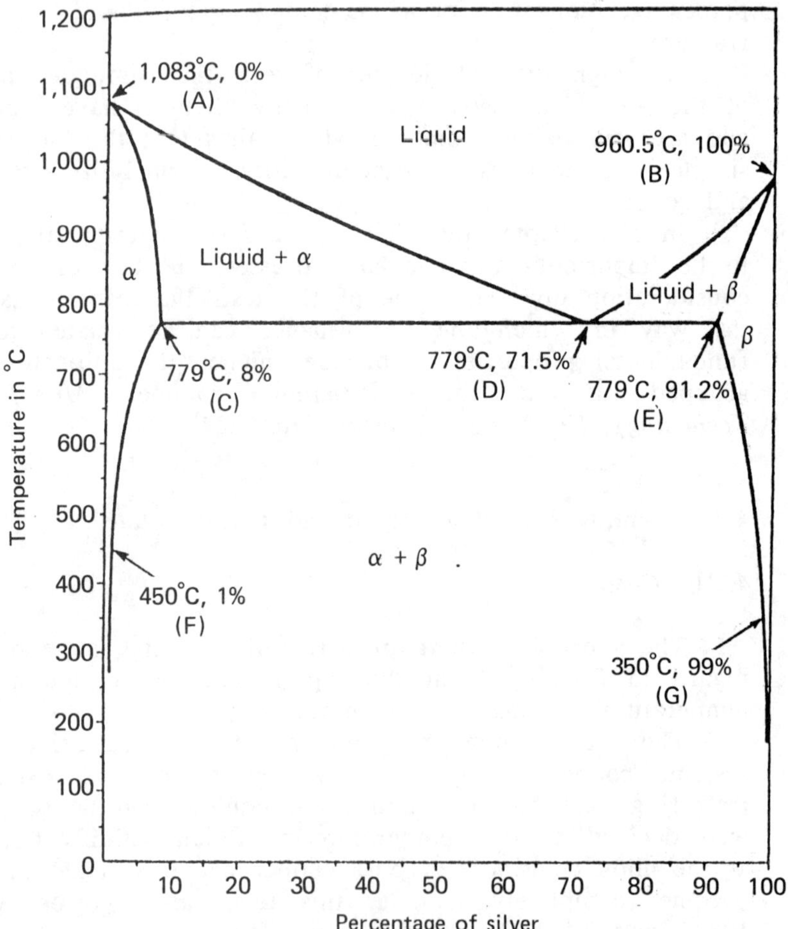

Figure 4.1 Equilibrium diagram for metals having limited mutual solid solubility (copper and silver)

The limited solubility of Ag in Cu gives the region α (8 per cent bounded by ACF) in which the behaviour resembles that of the single-phase Cu–Ni solid solution. The region ACD corresponds to liquid in equilibrium with this α-phase. At the high-silver end of the diagram there is a limited solubility region for copper in silver (up to 8.8 per cent).

Compositions between these limits will solidify to a mixture of α and β phases, either α or β having started to solidify first. The point D represents a special condition known as the eutectic, with a well-defined composition containing 71.5 per cent Ag. The two mutually insoluble phases, α and β, solidify simultaneously in fixed proportions and usually in very fine dispersion.

Other alloys, including the well-known example of carbon steel, also show eutectoid temperatures at which phases of a solid solution deposit simultaneously, but this does not directly affect the casting process. Some eutectoids may have special significance in producing superplastic properties. This can sometimes provide a competition for casting processes, for example in forming moulding dies[4].

A table of melting points and solidification ranges is given in Table 4.1. Melting-points give no real guide to the amount of heat which is required to melt a given weight of material. To determine this quantity, the specific heat of the solid and the latent heat of fusion must be taken into consideration. For specific heat, Dulong and Petit's law holds to within about 7 or 8 per cent. This states that atomic heat = specific heat $\times$ atomic weight = constant $\triangleq$ 25 J/g atom, and is sufficiently accurate for the present purpose. Furthermore, the slight increase of specific heat with increasing temperature may be neglected when assessing the quantity of heat required to melt unit weight of the material. This quantity of heat is also the amount which must be lost per unit weight of the casting in solidification and cooling, although the solidification time is governed, to a large extent, by the latent heat only.

This specific heat in the molten state governs the extent to which the molten metal is chilled as it passes through the gates and risers, i.e. the higher the specific heat, the less will be the chilling effect and consequently the smaller the difference between the lowest satisfactory pouring temperature and the melting-point temperature.

The thermal conductivity in the molten state has less effect than might be imagined at first sight, because the isothermals determined on the basis of conductivity alone are disturbed by the effects of turbulence and convection. During the cooling of the solidified casting, however, thermal conductivity plays a more important role in determining temperature gradients and consequently the degree to which internal stresses (due to differential contraction) are set up.

TABLE 4.1

Pure Metal	Melting Temperature (°C)
Iron	1535°
Copper	1083°
Aluminium	666°
Nickel	1455°
Zinc	419°
Lead	327°
Tin	232°
Magnesium	650°
Chromium	1850°

Alloy	Solidus-liquidus temperatures (°C)*
Stainless steel (18% Cr, 9% Ni)	1400°-1420°
Yellow brass (35/65)	950°-930°
Cartridge brass (30/70)	915°-955°
Red brass (15/85)	990°-1025°
Bronze (90% Cu)	1020°-1045°
Muntz metal	900°-905°
Naval brass (60/40)	885°-900°
Phosphor bronze	950°-1050°
30% Cupro-nickel	1170°-1240°
Aluminium-bronze	1050°-1060°
17S Aluminium alloy (0·75% Si, 4%Cu)	513°-641°
2S Aluminium alloy (1% Si, 0·2%Cu)	643°-657°
75S Aluminium alloy (0·5% Si, 1·6% Cu)	476°-638°
Cast Iconel	1370°-1400°
Cast Monel	1315°-1345°
Zamak 3 (4% Al)	380·6°-386·6°
Slush-casting zinc alloy (4}%Al)	380°-390°
Soft solder (20% Sn, 80%Pb)	183°-277°
Arsenical Babbitt metal	247°-353°

* (abstracted largely from *A.S.M.E.*
Metal Properties Handbook[5])

Since the flow of heat in a solidified casting is transient, rather than steady-state, it would be more logical to talk in terms of thermal diffusivity, particularly when making comparisons between different materials.

This can be seen from the basic heat-flow equation:-

$$\rho c \frac{\partial T}{\partial t} = K \nabla^2 T + Q \qquad (4.1)$$

where
ρ = density,
c = specific heat,
T = temperature (K),
K = thermal conductivity,
Q = heat produced per unit time in unit volume.

When applied to heat flow into a semi-infinite slab with uniform temperature applied to the face at $x=0$ this can be written:-

$$\frac{\partial T}{\partial t} = \kappa \frac{\partial^2 T}{\partial x^2} \qquad (4.2)$$

where $\kappa = K/\rho c$ is the thermal diffusivity. The same type of equation applies to heat conducted out of a large casting through a free surface.

The boundary conditions for the semi-infinite slab problem are defined as:-

$$T(x,0) = 0 \qquad (4.3)$$
$$T(0,t) = T, \ u(t) \qquad (4.4)$$

if we consider a unit step function $u(t)$. The solution is obtained using Laplace transform methods[6], as:-

$$T = T_1 \left[1 - \text{erf} \frac{x}{2\sqrt{(\kappa T)}} \right] \qquad (4.5)$$

The changes in temperature of a solidified casting can be important because of uneven contraction strains that may be produced, and also because of the dimensional difference between the pattern and the product. An allowance for the latter is often made by scaling the pattern with an appropriately enlarged pattern maker's ruler.

Apart from bismuth, antimony, and silicon, all pure metals contract during solidification, aluminium giving the largest magnitude (6 per cent). This solidification contraction of alloys is complicated by the fact that, since it takes place over a range of temperatures, there is always some 'solid' contraction of the phase which solidifies first. Thus, solid contraction figures for alloys tend to be higher than for pure metals.

It might appear, at first sight, that the amount of solidification contraction is too small to be of real importance. It should be remembered, however, that a casting solidifies from the outside towards the centre and, if there is no external feeding, the main loss of volume is felt near the centre. Since volume is proportional to the cube of the linear dimensions, a given loss in volume at the centre of the casting appears much worse than it would at the periphery, e.g. a 3 per cent contraction in volume on a 10 cm cube would change the linear dimensions by about 0.1 cm, if sustained at the outside, but would require a 'cubical void' of just over 3 cm side length at the centre (*sic!*)

The greater the range of temperature over which solidification takes place, the more difficult it is to ensure a sound casting. The alloys tend to freeze to equiaxed crystals but to form interconnected dendrites. The decreasing volume of liquid may not be able to penetrate between these arms, so small pores will be formed, especially near the heat centre of the casting where the channels are the smallest and the driving potential of the liquid is low. It is also well known to foundrymen that even an inexhaustible supply of liquid in the feeder will not prevent porosity in the final stages of solidification. Thermal diffusivity and the latent heat of solidification are also important features to be considered because of their influence on the temperature gradients and distribution of temperature. Complex shapes of casting introduce special problems. Some progress has been made in analysis of porosity formation in castings[7], and it may be hoped that finite element elastic-plastic thermal analysis will be helpful in the near future.

Hot tearing, where differential contraction or impeded contraction gives rise to large tensile stresses, may be particularly troublesome. The local strength at high temperatures is usually low but this is compensated to some degree by good ductility, except of course in hot-short alloys, where the grain boundaries of one of the phases may be brittle. Among the steels, hot tearing is more likely when the carbon content exceeds 0.1%, and is especially dangerous in the low-ductility region above 1300°C. Silicates, intergranular sulphides and other inclusions also increase the danger of hot tearing.

Alloys with long freezing ranges are also more susceptible to segregation of insoluble constituents or impurities. This can take two forms: (a) normal in which these are concentrated near the heat centre or in the uppermost part of the heat centre, and (b) _inverse_ in which a constituent is enriched at the surface. A classical case of inverse segregation is the formation of 'tin sweat' at the surface of a bronze.

Gravitational separation normally makes it impossible to cast combinations of metals of widely differing density, but spacecraft experiments have demonstrated that these can be produced (at a cost) in a non-gravitational environment.

The flow of molten metal in a mould is obviously of fundamental interest and, for complex castings with small-section channels, of practical importance too. It is thus disappointing that so little progress has been made in this field, although the difficulties that face either the analytical or experimental approach are evident. Experiments have been conducted with water flow in Perspex channels to identify turbulent flow areas, vortices and shut-off zones[8]. These are useful but cannot provide a complete analogy. More realistic patterns can be traced by using a radiographic filming technique during the pouring of real alloys into a two-dimensional channel system. Some measurements of coefficient of viscosity for molten metals have been made, so that it is possible to provide approximate allowances for changes from one metal to another in analogous gating systems[9].

The more general approach, however, has been to design special tests for evaluating the effect of viscous flow. Essentially, in all these tests, a specified amount of molten metal is poured into one end of a moulded channel and the length of the ensuing solidified 'rod' is taken as an index of the viscosity or fluidity of the metal. For practical convenience, the channel is frequently spiral in form so that a considerable length can be accommodated in a compact area.

Attempts to correlate such indices of fluidity with the coefficient of viscosity have never been successful, and this is understandable since the flow of a gradually solidifying mass along a channel is dependent on more factors than viscosity — indeed the problem may be more thermodynamic than hydrodynamic. It is probably better to regard such tests as providing empirical information which is of practical use. It is important to understand the way in which the test can be affected by the principal variables.

One of these is the temperature of the mould and, to illustrate this, Fig. 4.2 has been drawn from data given by KONDIC and KOZLOWSKI[10].

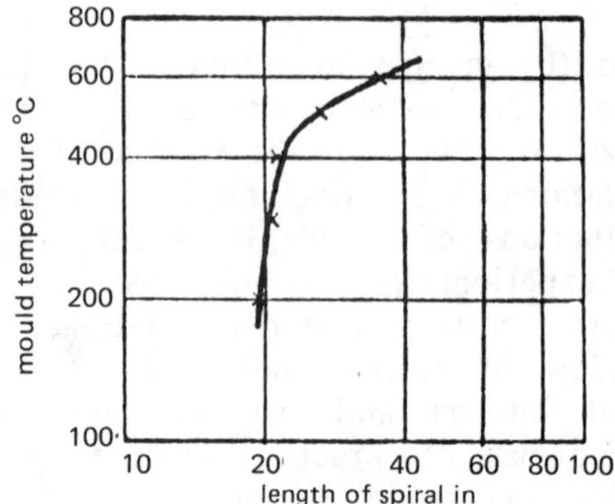

Fig. 4.2 Effect of mould temperature on
fluidity in a spiral channel test

Logarithmic co-ordinates were chosen so that it should be
clear that there was no power-law relationship between mould
temperature and solidified length; it is, nevertheless, clear (and
would be even clearer on linear coordinates) that the test is
relatively insensitive to mould temperature until the latter
exceeds 400 °C. The pouring temperature in these tests was
about 100C° higher than the melting-point, and this rather
high degree of superheat should be borne in mind when
drawing conclusions from Fig. 4.2. Under more practical
conditions, the mould temperature would have had to be held
below 300 °C in order to eliminate the effect of mould
temperature. It seems fairly well established that there is a
linear relationship between the fluidity index (solidified length)
and degree of superheat.

KONDIC and KOZLOWSKI[10] found that the head under
which the metal was poured had only slight effect − a 300
per cent increase in head (25 mm − 100 mm) caused the
solidified length to increase only from 135 mm to 160 mm.
This suggests, incidentally, that the head of metal in a riser
does not, *per se*, exert any great influence on the running of
a casting and that the difficulties experienced with thin
cross-sections are more attributable to the increasing
importance of surface tension as the sections become smaller.
Apart from the properties of the alloys themselves, the
properties of the mould materials have a profound influence
on the quantity or even feasibility of casting processes.
Chilling in metal moulds is an obvious example but even in
sand casting the use of zircon sand, together with other
process improvements, has greatly enhanced the accuracy of
light-alloy castings[11].

4.2.2 *Powder (Metallurgy and Ceramics), Compaction and Sintering*

Most if not all metals can be sintered, and many ceramics and cermets can be produced <u>only</u> by this method.

Metals such as tungsten and molybdenum have very high melting points and cannot be cast, but for others there may be special advantages in powder processing. Sintered aluminium powder (SAP), for example, makes use of the natural oxide skin on the small particles as a reinforcing constituent that enhances strength and creep resistance. Complex alloys such as the nickel-iron-molybdenum magnet steels can be produced without introducing undesirable phase changes. Modern techniques, including hot isostatic pressing (HIP), can produce components with very nearly full density[12]. As there is no liquid-solid transition, the dimensions can be very accurately reproduced and grain size can be controlled.

As seen in Chapter 2, an essential preliminary to the densification and sintering is a local welding at the points of contact between particles, followed by diffusion along the surfaces and in the bulk. The contact areas can be increased by pressure either at room temperature or during the heating. The latter is more effective but involves more expensive equipment.

In common with all diffusion-controlled processes, sintering proceeds much more rapidly at temperatures above about half the absolute melting point of the materials concerned. Ice, for example[13], can be successfully sintered in the temperature range of -25°C to -2°C; otherwise snowballs would need to be cast!

It might be thought that some consideration should be given to such matters as the effects of activation energy, sintering temperature in relation to melting temperature, etc. This, however, would be rather artificial since powder metallurgy techniques are principally used to obtain some specific property or to process a material which could not be processed in any other way (for technological or economic reasons), e.g. to obtain the porosity of an oil-impregnated bearing or to process highly refractory materials. In other words, the question of whether it is easier or more difficult to produce an alumina tool-tip than a tungsten carbide tip is immaterial once functional necessity has demanded alumina. In general, somewhat higher sintering temperatures are required for alumina than for tungsten carbide and it must thus be accepted that a higher temperature furnace may be required. Sintering times, optimum particle size, etc., may also be different but it is the functional demands, not the process, that dictate the material.

Nevertheless the economic balance of rival processes of casting, forging, sintering and machining may be very much influenced by the relative processing and product properties of the material required, for example in a gas turbine blade. These questions are considered in more detail than is possible here, by ALEXANDER[14].

4.3 Deformation Processes

4.3.1   *The 'Large-scale' Metal Forming Processes*

These are mainly processes such as hot forging, hot extrusion, and hot rolling, although there may be other rather specialized processes such as tube making which also fall into this category. The processes all involve predominantly compressive stressing of the material, and the deformation in forging by cogging is very similar to that which occurs in rolling.

In the hot working of metals, difficulties occur, some being associated with alloys having an intermediate constituent or phase which is brittle, or when the temperature is near the bottom of the range for hot working. Other problems which occur are typically:- formation of surface scale, inhomogeneous heating through the workpiece, local melting, difficulty of lubricating at very high temperatures, decarburization of steels, carbon pick-up, to name but a few. Thus terms such as 'forgeability', 'rollability', 'extrudability', rarely occur for hot working and few special tests have been developed. To assess forgeability, for example, the most prominent tests are the 'single blow drop test' and the 'hot twist test', mentioned by KOBAYASHI and DODD[15]. In the first of these a specimen is held at a given temperature for ten minutes, then subjected to blows of different impact intensity and the resulting percentage reduction in height measured. It is found that there is a relationship between impact energy E and percentage reduction in height ΔH of the form:-

$$E = b.\Delta H^{(1.56-0.47\ \ln\ b)} \tag{4.6}$$

in which b is a constant for a given metal. If $\ln\ \Delta H$ is plotted against ln E a straight line results and the slope may be taken as an 'index of forgeability' or 'malleability number'.

In the hot twist test a 14.3 mm (9/16 in) diameter bar 559 mm (22 in) long is tested, the object being to discover the best forging temperature. The bar is held between the chucks of a lathe, with its central portion inside a furnace. The sample is heated to the test temperature and one end rotated until failure occurs. The number of twists is counted automatically. Although highly empirical, it is claimed that both these tests give a reliable indication of forgeability. Brasses, especially leaded brasses, are prone to brittleness in the hot-working range and edge cracking easily occurs in hot rolling.

There is some hope that finite element analysis will allow the prediction of such conditions for failure[16], but there is at present no satisfactory simulative test. Hot upsetting of a cylindrical billet will reveal any tendency to crack at the centre or on the periphery. The form of such cracks has been studied[17] but direct correlation with practice is difficult.

Useful information relating to the yield strength of metals can be obtained in hot high speed compression tests. When conducted at various temperatures and strain rates, these provide basic stress strain values for use in theoretical analyses and load calculations. At very high strain rate a torsion test of the Hopkinson bar type may be used, and it has even been suggested that machining tests can be interpreted to provide yield stress data.

Unfortunately, a manufacturer is usually more interested in knowing whether the material can be forged, rolled or extruded without it breaking up. This important information cannot easily be obtained from any simple tests, and often requires a close simulation of the actual process, if not even a trial run.

4.3.2 *The 'Small-scale' Metal Forming Processes*

Similar considerations to those just discussed apply to the assessment of metal for use in cold forging, cold extrusion, cold rolling, wire drawing, and tube sinking operations. The effects of temperature and rate of straining are virtually non-existent in cold working operations so that compression tests are much easier to carry out. Lubrication of the tool-specimen interface presents no problem, and incremental loading can be used. To estimate the loads required in any cold working process the basic stress strain curve must be known, and this can be found by any of the methods described in Chapter 1. Failure of material subjected to these operations is usually due to fracture, often caused by secondary tensile stresses in predominantly compressive operations. In recent years it has been recognized that these cracks sometimes follow the directions of maximum shear, as deduced from slip line fields, and are then probably the result of intense localized shear[18].

There is a broad correlation between the propensity for cracking in a forming operation and the ductility, measured as percentage elongation or percentage reduction in area of a tensile test. However, as already mentioned, these two measurements do not always tally and their translation to extrusion, rolling and drawing is not reliable.

Sheet forming is a very important process, largely governed by uniaxial or biaxial tensile forces. Various material property tests have been proposed[19], among which the most important are various bend tests and cupping tests.

4.3.2.1 *Bend tests*

A strip is cut from the sheet and bent over a specific radius. Sometimes loads and deflections are recorded but more commonly the strip is bent through 90° in the jaws of a vice and then through 180° in the reverse direction. The number of times that this can be done is an indication of crack resistance. A similar test, often used for sheet destined for car manufacture, bends the strip back upon itself, to make a rounded edge. No cracks should appear, although the small radius makes this a severe test.

4.3.2.2 *Cupping tests*

This generic term is used to cover a whole range of tests for the 'formability' of sheet. A punch is pushed into a sheet, which is supported on a die and clamped by a blank holder; the punch forms a bulge if the sheet is of large area, or a cup if the sheet is in the form of a circular blank. The term is unfortunate since it seems to refer to a test in which a cup is formed, whereas this is not always the case. Cupping tests are much more important and useful than the previous tests mentioned, and will therefore be discussed in some detail.

The most popular of the many cupping tests which have been developed are the Swift Test and the Erichsen Test, with the Olsen Test following close behind. In the Swift Test a flat-bottomed cup is drawn from a circular blank the dimensions of the tools being within certain specified limits (KEMMIS[20]). A number of blanks of slightly different diameters are drawn to determine the blank diameter above which all cups will fracture. This diameter of blank divided by the punch diameter then defines the 'limiting drawing ratio', or 'LDR'.

Useful though this test is, it will obviously not predict the behaviour of a given sheet of metal when subjected to *any* pressing operation. It will certainly predict the behaviour of a sheet when subjected to the operation of deep drawing a cup of the same size as that given in the specification. It will also give a good indication for deep drawing a large cup, although the sheet will be thinner in relation to the large cup than it was to the test cup, so that some difference *must* result.

In a practical pressing operation, however, the part to be formed is usually of complicated shape, and it is difficult to say how much 'stretching' or how much 'deep drawing' will take place. In the Swift Test the maximum amount of deep drawing is impressed on the sheet so that if little or no deep drawing occurs in the process of interest the Swift Test will be of little use.

The Erichsen Test, on the other hand, is a test in which the deformation is predominantly by stretching. The specimen is in the form of a wide (90 mm) strip which is clamped against the die (27 mm diameter), and a punch of spherical shape (20 mm diameter) is pressed against the sheet. Because of the width of the strip and the clamping force, 'drawing-in' of the metal is prevented almost completely, so that a bulge is formed in the sheet due to the sheet being stretched over the punch. The test is shown diagrammatically in Fig. 4.3, with the dimensions specified in the provisional Continental specification. The punch is pressed into the sheet until fracture occurs, at which point the test is stopped immediately and the depth of the bulge noted. This depth (mm) gives the *Erichsen number*.

The Erichsen number obviously gives a measure of ductility of the sheet in the plane of drawing *under biaxial stress conditions*. This is information which cannot be derived directly from a uniaxial tensile test. The Erichsen Test is therefore useful for assessing the ability of a metal to stretch in biaxial stressing, and also for revealing the grain structure and directionality or texture of the sheet.

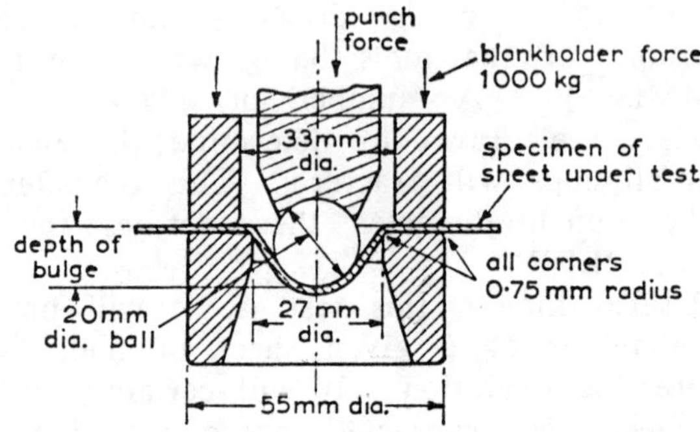

Fig. 4.3 The Erichsen Test (D.I.N. 50101)

If the sheet is very anisotropic so that its properties in the plane of the sheet differ with orientation, the crack at fracture will appear as a straight slit across the bulge. In an isotropic sheet in which there was no variation of properties with orientation in the plane of the sheet, then the fracture occurs symmetrically around the bulge.

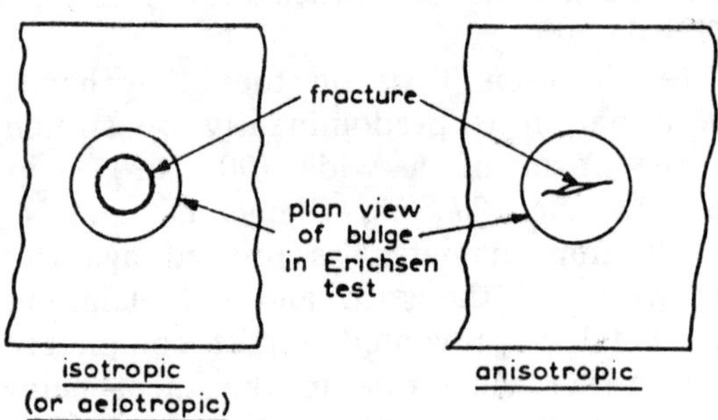

Fig. 4.4 Appearance of fracture in the
Erichsen Test specimen

These effects are illustrated in Fig. 4.4. Also, if the grain size is too large the stretching of the metal over the spherical punch will give rise to the formation of a rough surface resembling that of 'orange peel'. The rougher the surface the larger the grain size and, as shown by JEVONS[19], it is relatively easy to distinguish between average grain sizes of 0.014, 0.045, and the 0.20 mm, simply from the visual appearance of the surface of the bulge.

The Erichsen Test is one of the most convenient and useful tests in existence for the routine inspection of sheet metal. It must be carefully controlled, however, since there are several sources of error which can be introduced, giving scatter in the resultant Erichsen numbers, as shown by KAFTANOGLU and ALEXANDER[21].

The Olsen Test differs little from the Erichsen Test; in fact the diameter of the die is 50 mm instead of 27 mm. It is possible to draw up a conversion table of Olsen to Erichsen numbers although the inevitable scatter that exists makes this rather a doubtful procedure. The Olsen Test has received more consideration in the U.S.A.

A further variant is the hydraulic bulge test in which the sheet is clamped over an orifice so that an approximately hemispherical bulge can be produced by applying hydraulic pressure to one side. This avoids friction on the punch and is a direct measure of the material property, though there are problems in relating this to practical conditions in which friction is present.

4.3.2.3 R-value tests

An important feature of most sheet metal is that, because it is produced by longitudinal rolling, it has a pronounced anisotropy in its structure. This influences the tensile ductility and also the tendency to produce ears on a drawn cup[22] usually four in number but sometimes six. These are waves around the edge of the cup as sketched in Fig. 4.5, which have to be cut off and so waste material.

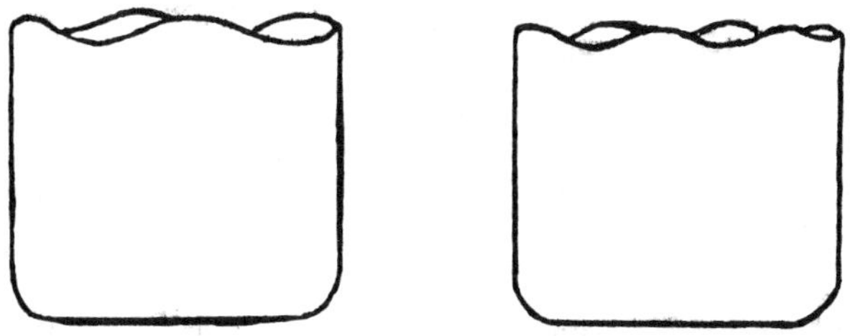

Fig. 4.5 Earing of cups from a deep-drawing test.

The anisotropy can be assessed in a tensile test by measuring the r-value, defined as :-

$$r = \frac{\text{width strain}}{\text{through-thickness strain}}$$

High r-values (typically from 2 to 6 for titanium sheet) mean that the thickness is reduced less in proportion to the width reduction and consequently the material will better withstand the thinning that occurs during bending under tension.

Other proposals have been made, for example that the strain hardening rate should be included in the assessment[23,24], since the tensile instability initiating necking depends upon this parameter.

4.4 The Machining Processes

In a technological world where other processes (forming processes in particular) exert a continuously increasing challenge to machining, the latter process has several advantages which, in certain circumstances, can be very persuasive, viz.:-

(1)   the machines used for metal removal are comparatively cheap, operate at low power levels, and are versatile;
(2)   elaborate form tools and dies are not necessary;
(3)   tool wear is not an expensive item in most cases;
(4)   the process is almost independent of the material of which the component is to be produced;
(5)   there are many parameters which can be varied to meet economic and technical difficulties.

It may be seen that machining is most advantageous in jobbing and small production work and most vulnerable in large-scale production where the high capital cost of forming machines and heavy expenditure on dies can be shared among tens of thousands of components.

There are, of course, components which can be produced by effectively only one method (e.g. porous bearings, thin walled components such as toothpaste tubes) but, in the majority of cases, there are several methods capable of producing a given component, and economics becomes a governing factor. · It is important for the user to know the following information, as far as possible:-

(1)   how long the cutting tool will last before performance becomes unacceptable, in other words how many components can be made without changing the tool;
(2)   the maximum rate at which metal can be removed while

retaining a specified tool life;

(3)   the forces acting on the tool, which will determine whether chatter occurs on a given machine and how well dimensional tolerances can be held. These forces affect tool wear, heating and failure;

(4)   the surface finish that can be obtained within the above constraints and at an economic rate of production;

(5)   the form of the chip, in small easily-disposable fragments, or long and tangled-up pieces of swarf.

Although these very different criteria impose different requirements, and each will vary according to speed, feed and tool geometry, there are still attempts to define 'machinability' or a 'machinability index' as a material property. It is true that a leaded brass will in general be easier to machine than the straight Cu-Zn alloy, and resulphurizing undoubtedly improves the cutting of mild steel, but in more detail there is no substitute for actual or somewhat simplified cutting trials. Fortunately lathe tests, drilling, milling or tapping are quite easily carried out, but a considerable quantity of material may be consumed in high speed cutting tests and apparently minor variability in material within specification tolerances is a notorious cause of unreliable results. It is by no means easy to specify optimum conditions for cutting a given alloy, though handbook recommendations do exist, and there is still no reliable theory of machining, even for the simplest conditions. Some guidance can be obtained from elementary analysis, but only in very general terms.

It was therefore accepted long ago that cutting tests are necessary, and the classic studies of TAYLOR[25] set the pattern in 1907 for systematic variation of speed, feed, depth of cut, tool geometry and lubricants. The famous Taylor equation:-

$$VT^n = C \qquad (4.7)$$

is still widely used to relate the cutting time T to produce a standard flank wear land (e.g. 0.75 mm) to the speed V. The two constants C and n are specific to the material and cutting conditions.

This equation is remarkably well followed under a wide range of conditions. It is common practice to plot the logarithmic form:-

$$\log V = \log C - n \log T \qquad (4.8)$$

to obtain an approximately straight line, from which the speed to give a tool life of 60 minutes ($V60$) or sometimes 30 minutes ($V30$) is deduced, but this equation applies only to high speed cutting. The implication that tool life becomes infinite at very low speeds is far from true. Partly because of strain hardening or built-up edge formation or other factors, the cutting forces may rise dramatically at low speeds. It should be recognized also that in many practical instances, the alloy can be cut very satisfactorily at much greater speeds than are actually used, especially when carbide or ceramic tools are used.

Tool life, as indicated above, is by no means the only criterion. Further information on surface finish, chatter and accuracy can often be obtained by short time tests that may last as little as one minute. These establish steady conditions but use relatively little material. Such tests are useful for assessing appropriate speed, feed and angle settings[26]. Despite the difficulties of machinability testing, it is possible to make general comments about the major groups of material as follows.

4.4.1 *Aluminium*

Excellent tool life is obtained with HSS (high-speed steel) tools because the temperatures are limited by the melting point of aluminium, ($\simeq 659°C$) or its alloys. Speeds up to 300 m/min are used, but Al-Si castings provide an exception because the large Si grains can cause severe flank wear. These alloys are best machined with diamond.

Thick continuous chips are formed when cutting pure aluminium, which may clog the tools, especially drills and taps. Some alloys, including "Duralumin" (a trade name for the aluminium alloy containing approximately $4\frac{1}{2}$ per cent Cu), form a built-up edge at low cutting speeds.

4.4.2 *Copper Alloys*

Pure copper (Oxygen-Free, High-Conductivity $\equiv$ OFHC), like most unalloyed metals, produces thick chips and requires high forces at low cutting speeds, but the performance improves at speeds above about 100 m/min. Cold working reduces the effective strain hardening rate and thereby also thins the chips and reduces the forces. With low cutting speed and annealed stock, the surface finish on the machined workpiece is poor.

Alloying greatly improves the surface finish and reduces the forces. The most popular Cu-Zn alloy, 60/40 brass, contains both α and β phases and is well known to possess good machinability, by whatever criterion that is judged. Nevertheless the swarf produced is continuous, so chip disposal is facilitated by the addition of lead at 2-3 per cent by weight in the alloy, which gives short broken chips. This also reduces the forces and the surface finish is further improved. Other 'free-machining' additions such as S or Te can be used. High speeds should be used to avoid built-up edge formation with the basically two-phase alloys. 200 m/min is often used on small diameter stock but higher speeds are possible. 70/30 brass contains only the α phase. This requires slightly higher forces than 60/40 at high speeds but forms no built-up edge and can be cut easily at very low speeds if required.

4.4.3 Cast Irons

Pure iron (ARMCO) has very poor machining quality. It forms thick chips but no built-up edge.

Flake graphite cast irons have good machinability, by all criteria, attributable to the graphite acting as a free-machining agent, though it is now believed[27] that it does not act directly as a tool-face lubricant. Ferritic or pearlitic cast irons can be machined dry but cemented carbide tools are recommended since the temperatures may be high.

Highly alloyed cast irons and chilled irons contain very little graphite but contain Fe_3C and other hard carbides which make machining very difficult, with very severe wear problems. Even with tungsten carbide tools the cutting speed may be only about 5 m/min.

4.4.4 Carbon Steels

Low-carbon steels are ductile and do not machine well, especially in the annealed state, but the forces are less than with pure iron and the chips are thinner. High speeds are recommended.

Increasing carbon content gives smaller tool-face contact, less built-up edge and lower forces, but the hard iron carbide increases wear. The best machinability is obtained by spheroidizing the cementite but this is a slow and expensive process. Cutting speeds are restricted by the high forces and high temperatures generated, but about 300 m/min can be used with carbide tools.   At about 5 m/min there is a strong built-up edge formation, as shown in Fig. 4.6(c).

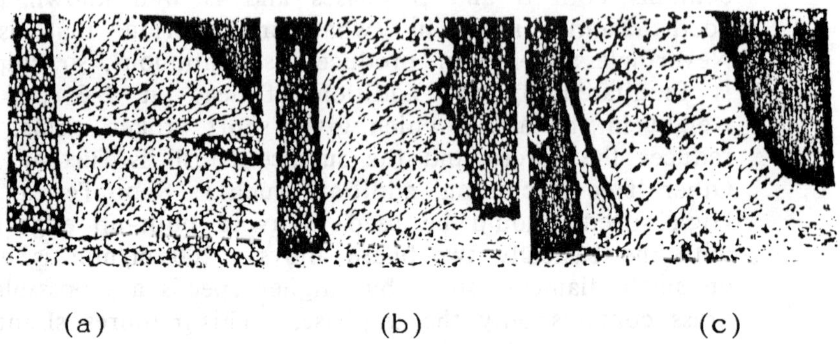

(a)   (b)   (c)

Fig. 4.6 Chip forms produced in low-carbon steel

(a) annealed (b) work hardened (c) in the built-up-edge region

Free-cutting steels give longer tool life, better surface finish, reduced tool forces and power consumption.   They also permit higher speeds and produce small broken chips that are easy to handle.   They give much more reproducible cutting than other low-carbon steels and are widely used in automatic lathes.   These alloys usually contain 0.15-0.44 per cent C, with MnS as the effective additive.   Lead is also sometimes used.

Because iron tends to form an alloy with WC-Co, steel cutting grades of cemented carbides usually contain TiC or TaC additions to reduce wear.

4.4.5 *Alloy Steels*

Some alloying elements improve machinability by reducing the ductility of the iron, but wear is generally increased, especially by alloying agents that form hard carbides.

Austenitic stainless steels are particularly difficult to machine.   The forces do not differ greatly from those for carbon steels at comparable speeds, but the stainless steel tends to bond strongly to HSS or WC-Co tools.   The temperatures are somewhat higher and wear occurs by rake-face cratering.   These alloys strain harden rapidly and a high feed-rate is recommended to avoid the effect of the hardened layer.

Free-cutting stainless steels are available, containing either sulphur or selenium. The latter are preferred because the corrosion resistance is reported to be less impaired.

Other austenitic steels can best be machined at elevated temperature, preferably by heating the cutting zone just ahead of the tool.

4.4.6 *Nickel Alloys*

Nickel and all its alloys are difficult to machine, although the melting point of Ni (1452°C) is lower than that of iron (1535°C). Tool life tends to be short and failure occurs by deformation and flank wear at low cutting speeds. 50 m/min is suitable for commercially pure nickel when high-speed steel tools are used.

Alloying elements may increase the strength of nickel. They usually reduce the cutting forces but tool wear is increased and the maximum removal rate reduced.

Creep-resistant aerospace alloys are very difficult to cut or grind, partly because of their specific combination of high strength and temperature resistance.

Cubic boron nitride (CBN) coatings on tungsten carbide tools have been recommended for machining nickel alloys, but they are very expensive. An alternative may be local heating or 'hot machining'.

4.4.7 *Titanium Alloys*

Titanium and its alloys present many problems, though the cutting forces are lower than for iron and nickel alloys of comparable hardness. This may be associated with the relatively small tool contact area developed and the consequent high stresses.

Straight WC-Co cutting tools have longer life with titanium alloys than the steel-cutting WC-TiC-Co or WC-TaC-Co grades, but speeds may still be limited to 30 m/min.

4.4.8 *Coolants*

Tool temperature is a very important feature in machining. The main advance due to the introduction of HSS tools was the ability to withstand higher temperatures. Any method of cooling the tool will usually improve tool life. If in addition some active ingredient in the coolant reduces the tool-face contact, the forces will be reduced and tool life further increased, usually with a better workpiece surface being produced. Sometimes, however, the reduction or prevention of built-up edge formation by a coolant can result in greater rake-face wear.

4.5 Welding

Weldability, like the other related general terms, is somewhat vague. It has been suggested by ROLLASON[28] that three separate aspects should be considered: mechanical tests on sound welds, metallurgical study of the heat-affected zone, and crack sensitivity.

Fusion welding has some resemblance to casting. In general the higher the melting temperature the more difficult the process becomes. The high rate of cooling for welds, and the steep thermal gradients introduce additional problems. Trapping of gases, evolved because their solubility in solid metal is less than in the liquid, becomes an even more serious problem in welds. It is therefore common to use oxygen-removing elements such as manganese or silicon in the slag, together with titanium to combine with nitrogen. An alternative is to weld under inert or reducing gas.

Other specific problems arise with certain metals, such as the high thermal diffusivity of copper, which causes difficulty in introducing sufficient heat without local burning.

4.5.1 *Mechanical Tests*

The strength of sound joints can be assessed by various tensile or shear methods applied to butt joints or other configurations. These range from measurements on carefully prepared tensile specimens to judicious blows on a fillet weld with a heavy hammer.

An important part of any weld test should be either a thorough non-destructive examination by ultrasonics, radiography or some other suitable technique, before the test; or a macroscopic examination of the fracture afterwards, to ensure that the weld was indeed sound.

Dynamic testing is desirable for welded joints since fatigue resistance can be very seriously reduced by welding. Specifications for the maximum stresses allowed for welded joints of various sorts under cyclic loading have been established by TIPPER[29]. Figure 4.7 shows a range of joints. Ideally, the welded structure should have the same strength as a monobloc construction, which provides a standard of comparison.

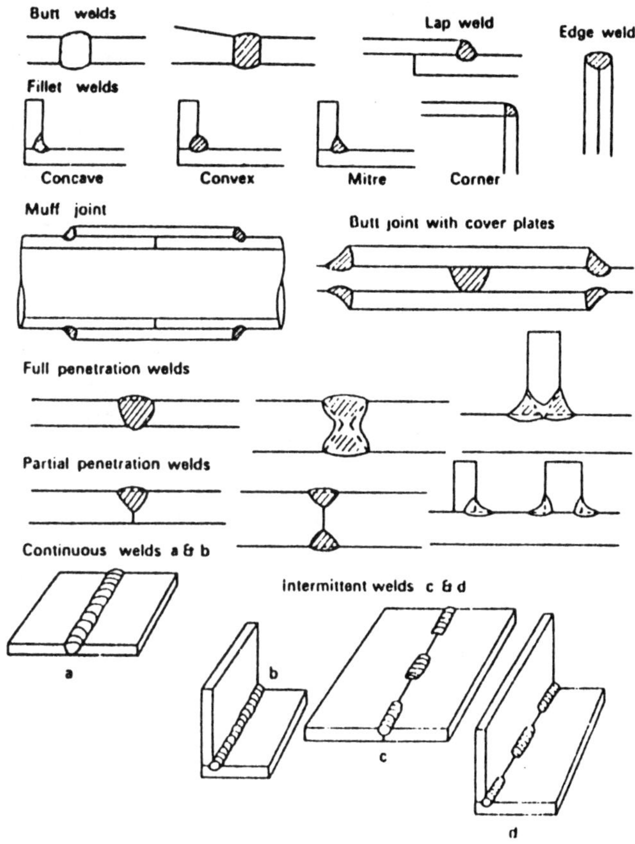

Fig. 4.7 Types of arc-welded joints
(from MILNER and APPS[30])

4.5.2 *Metallurgical Examination*

A metallographic section of a weld zone (Fig. 4.8) often shows a wide range of structures, in fact a microcosm of almost the entire range of heat-treatment of the alloy (see ROLLASON[28]). Quenched, tempered and annealed grains can be identified, especially in multipass welds. For critical applications such as reactor pressure vessels, careful post-heat-treatment is essential.

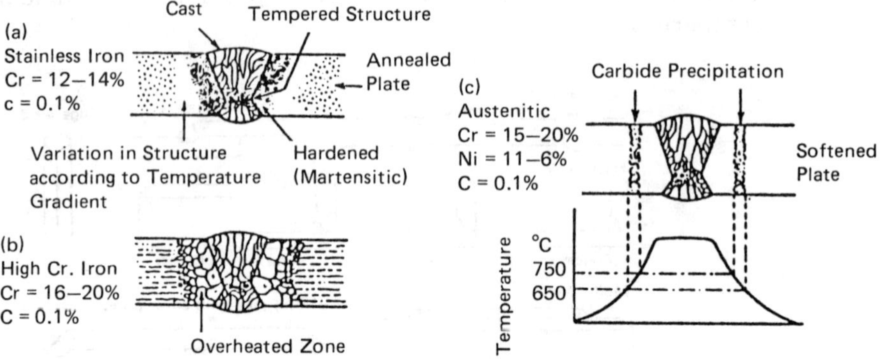

Fig. 4.8   (a) A cross-section of a weld in stainless iron,
etched to show the various microstructures,
(b) high chromium iron, (c) austenitic stainless steel

Microhardness   traverses   {Fig.   4.9(a)}   can   also   be
informative but are time-consuming.   The maximum hardness
value can be critical and preheating may be essential.   As Fig.
4.9(b) shows, larger electrodes require less preheating of the
workpiece[31].

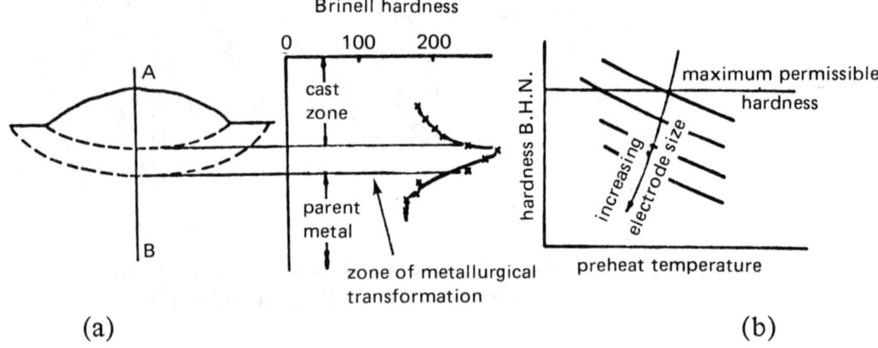

Fig. 4.9 (a) The variation of hardness
with depth in a weld nugget.
(b) Effect of preheat temperature on hardness
for various electrode sizes

4.5.3 Crack Sensitivity

Welded structures are particularly susceptible to cracking
especially in steels that are notch-sensitive (see TIPPER[29]).
Simple lap or T joints can be used for tests and various

configurations have been adopted as standards, some of which are shown in Fig. 4.10.

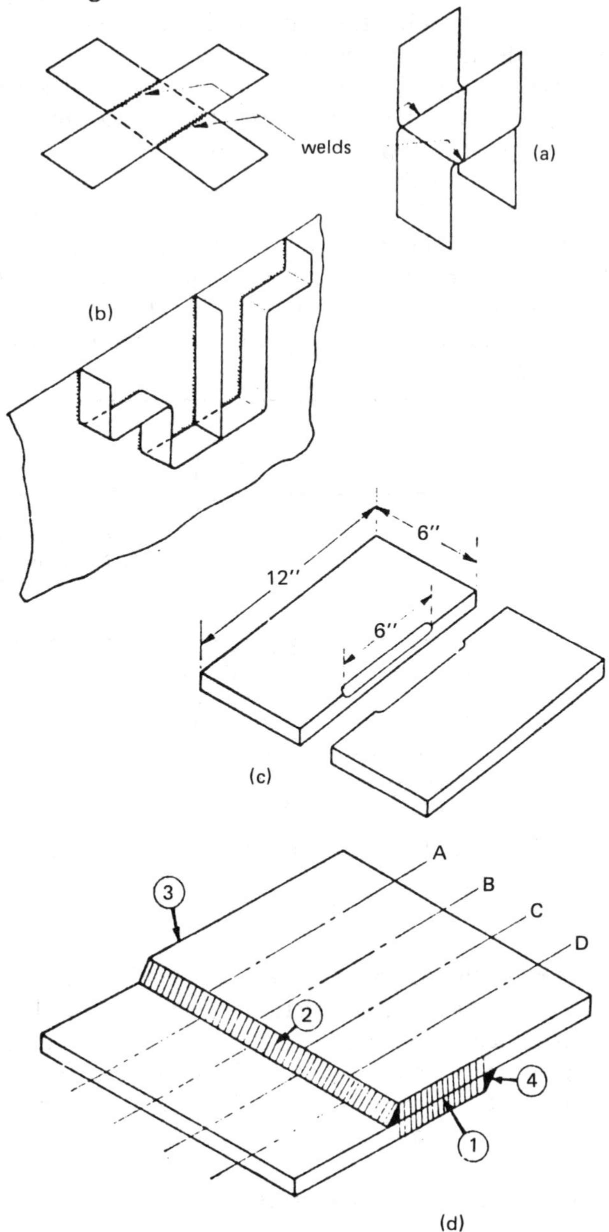

Fig. 4.10 Various weldability test pieces:- (a) Cross Test (b) Krupp Test (c) Services Test (d) O'Neill Test

These are examined visually or by NDT for cracking. Other tests use square or circular welds on the edges of a plate or disc so that self-stressing is introduced as the first sections cool while the remainder is being made.

Deliberate notches can be cut with a small radius or a V shape, to introduce a stress concentration. Fig. 4.11 shows two different types of specimen for the little known but effective "Schnadt" test (the upper specimen being un-notched) both suitable for use in Charpy impact testers. The notch sensitivity of the material itself can be important, since failure of a structure may not necessarily start in the weld region. The susceptibility of steels to brittle fracture at various temperatures has been extensively studied with notched tensile specimens.

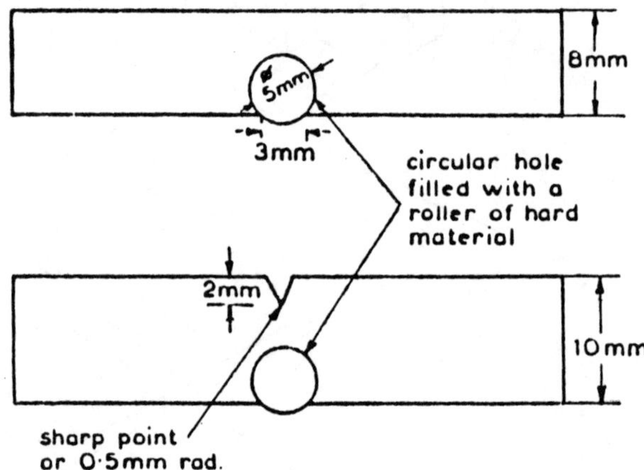

Fig. 4.11 Schnadt specimen for Charpy impact testing

4.5.4 *Resistance and Spot Welding*

From the point of view of testing, resistance welding has many of the characteristics of fusion welding by arc, gas, electron beam or laser techniques. Steep temperature gradients are produced and a range of microstructures can be observed. Since most of the joints will be of lap or butt type, tearing or tensile tests are commonly used.

4.5.5 *Weldability of Common Alloys*

Some general comments can be made, subject to the recognition that weldability has no clear definition.

4.5.5.1 *Copper*

This metal welds easily, provided that enough heat can be provided to overcome the high conductivity. Preheating and insulating may be needed. Also there is a low ductility in copper at 600°C, so stressing at a dull red heat must be avoided. This is particularly evident in tough pitch copper.

4.5.5.2 *Brass*

Brasses can be welded satisfactorily but volatilization of zinc causes problems. Welds should be annealed to prevent season cracking (stress corrosion failure).

4.5.5.3 *Bronze*

Bronzes have a wide solidification range, which can lead to porosity and cracking. Aluminium bronzes can be welded with a metallic arc but it is difficult to avoid oxide films in the weld, which reduce ductility. Tin or aluminium bronze electrodes are recommended for copper arc welding.

4.5.5.4 *Nickel alloys*

Nickel alloys exhibit hot shortness at about 700°C, so stressing in this temperature region should be avoided. Deoxidizers in the flux are required, but silicon promotes cracking and should be avoided.

4.5.5.5 *Aluminium alloys*

Aluminium-silicon alloys can be welded readily, but pure aluminium requires an active flux to remove the strongly adherent oxide. Duralumin type alloys tend to have low weld strength.

4.5.5.6 *Steel and its alloys*

Arc welding is very widely used for steels, and sound welds can be produced with normal care. It is essential to remove the flux thoroughly before running a second pass. The quality of the weld depends very much on the skill and conscience of the welder. Where the steel may be susceptible to brittle fracture, a low hydrogen flux is essential unless CO_2 or inert gas shielding is used. Welded structures should whenever possible be stress-relieved at about 600°C, even by a hand-held torch, to overcome residual stress problems and temper brittleness. Pre-heating as well as post-treatment is recommended for large structures. Alloy steels and precipitation hardening alloys may require specific post-heat-treatment to restore their properties in the weld and heat affected zone. Stainless steels have a dual problem. The tenacious oxide film may be removed but carbon must be avoided since brittle chromium carbides may form which also reduce corrosion resistance. Nevertheless austenitic 18/8 stainless steels can be arc welded very satisfactorily if titanium, niobium or other stabilizer is incorporated in the alloy. This helps to prevent weld decay due to chromium depletion by carbide formation.

REFERENCES

(1) ALEXANDER, J.M., BREWER, R.C. and ROWE, G.W., *Manufacturing Technology, Vol.2: Engineering Processes.* Ellis Horwood Ltd., 1986.

(2) FLEMINGS, M.C., *Solidification Processing.* McGraw-Hill, 1974.

(3) TWEEDDALE, J.G., *Metallurgical Principles for Engineers.* Iliffe Books Ltd., 1962.

(4) ROWE, G.W., *"Superplasticity" in Metal Forming and Impact Mechanics* (Ed. S.R. Reid). Pergamon Press, 1985.

(5) *Metal Properties Handbook.* A.S.M.E.

(6) CARSLAW, H.S. and JAEGER, J.C., *Conduction of Heat in Solids.* Oxford University Press, 1959.

(7) MURPHY, A.J. (Ed.), *Non-ferrous Foundry Metallurgy.* Pergamon Press, London, 1954.

(8) KONDIC, V., *Metallurgical Principles of Founding.* Edward Arnold, London, 1968.

(9) WIESER, P.F. (Ed.), *Steel Castings Handbook.* Steel Founders Soc. Amer., (5th Ed.), 1980

(10) KONDIC, V. and KOZLOWSKI, H.J., "Fundamental Characteristics of Casting Fluidity", *J. Inst. Metals* 1949,7,665.

(11) CAMPBELL, J. and WILKINS, P.S.A., "New Developments in Light Alloy Founding", *Brit. Foundryman* 1982,**75**,233-236.

(12) HAUSNER, H.H. and MAL, M.K., *Handbook of Powder Metallurgy*. Chem. Publ. Co., New York, 1982.

(13) KINGERY, W.D., "Regelation, Surface Diffusion and Ice Sintering", *J.App.Phys.* 1960,**31**,833.

(14) ALEXANDER, J.M., "The Economics of Alternative Methods of Metal Processing", *C.I.R.P. Annals* 1984,**33**(2),

(15) KOBAYASHI, H. and DODD, B., "A Review of Dynamic Torsion Testing Techniques" (in press).

(16) CLIFT, S.E., HARTLEY, P., STURGESS, C.E.N. and ROWE, G.W., "Fracture Initiation in Plane Strain Forging", *Proc. 25th Int.M.T.D.R.Conf. Birmingham* 1985,413-419.

(17) KOBAYASHI, S., "Metalforming and the Finite-Element Method − Past and Future", *Ibid.* 17-32.

(18) DODD, B., "Shear Instabilities in Blanking and Related Processes", *Metals Technology* 1983,**10**,57-60.

(19) JEVONS, J.D., *The Metallurgy of Deep Drawing and Pressing*. Chapman and Hall, London, 1941.

(20) KEMMIS, O.H., "The Assessment of the Drawing and Forming Qualities of Sheet Metal by the Swift Cup-Forming Test", *Sheet Metal Industries* 1957,**34**,203,251,

(21) KAFTANOGLU, B. and ALEXANDER, J.M., "An Investigation of the Erichsen Test", *J. Inst. Metals* 1962,**90**,457.

(22) HILL, R., *The Mathematical Theory of Plasticity*. Clarendon Press, Oxford, 1950.

(23) WHITELEY, R.L., WISE, D.E. and BLICKWEDE, D.J., "Anisotropy as an Asset for Good Drawability", *Sheet Metal Industries* 1961,**38**(409),349.

(24) WALLACE, J.F., "The Assessment of the Formability of Sheet Metal by Cupping Tests and from the Stress-Strain Characteristic", "Deep Drawing", *I.D.D.R.G. Paris Conf.* 1960.

(25) TAYLOR, F.W., "On the Art of Cutting Metals", *Trans.A.S.M.E.* 1906,**28**,31-350.

(26) TRENT, E.M., *Metal Cutting*. Butterworths, London, (2nd Ed.), 1984.

(27) CHILDS, T.H.C. and ROWE, G.W., "Physics in Metal Cutting", Rep.Prog.Phys. 1973,**36**,223-288.

(28) ROLLASON, E.C., *Metallurgy for Engineers*. Edward Arnold, London, (3rd Ed.), 1961.

(29) TIPPER, C.F., *The Brittle Fracture Story*.
Cambridge University Press, 1962.
(30) MILNER, D.R. and APPS, R.L., *Introduction to Welding and Brazing*. Pergamon Press, 1968.
(31) SÉFÉRIAN, D. *Métallurgie de la Soudure*.
Dunod, Paris, 1959.

Chapter 5

THE EFFECT OF PRODUCTION PROCESSES ON THE FUNCTIONING OF MATERIALS

5.1 Introduction

The methods by which an engineering component can be made are generally quite numerous. The final choice will always be dictated either directly or indirectly by economic considerations, which are not always apparent at the outset. The overall process of designing an article which will perform satisfactorily the functions desired for a specified time is very complex, involving optimization on an economic basis either by intuition or 'common-sense' methods, or by mathematical analysis. All the necessary information for this is often not available in the design stage, so that past experience and/or 'cut and try' methods must be used. The choice of production processes to be applied depends upon various considerations, in that such factors as the availability of the necessary equipment, cost of the process and the effect of the process on the material will all figure in the economics of production.

Many of the processes involve intermediate heat-treatments such as preheating, annealing and inter-annealing and the effects of such metallurgical treatments in the present context will be referred to in a separate section, since they involve considerations additional to those already mentioned in Chapter 2. Many of the effects of the various processes on metals are detrimental, and a very complete account of the possible defects and failures has been given by POLUSHKIN[1], together with a description of their origin and methods of elimination.

In a final discussion are compared several different processes for producing a component from this point of view, to illustrate the considerations involved.

5.2 Forming from the Liquid State

Most of the principal defects which may occur in a casting originate either at or after the point when the flow of metal is appreciably hindered by solidification. Thus, as before, it is necessary to consider pure metals and skin-forming alloys separately from those alloys which have a long freezing range.

It is essential that castings which freeze by skin formation should solidify continuously towards the feeders in such a manner that no completely liquid section is cut off from the feeders by a 'bridge' of solidified metal. This has been called 'controlled directional solidification' and implies a control over the ratio of longitudinal solidification (parallel to the mould walls) to lateral solidification (perpendicular to the mould walls).

The only way in which this control can be effected is to select the mould material and the gating, feeding, and pouring methods in such a way that the temperature gradients will ensure continuous solidification towards the feeder(s). It is assumed here that two other important factors, viz. the shape of the casting and the material from which it is to be cast, have been selected and cannot be adjusted, although it should be clear that an alteration in the shape of a casting may exert a very profound effect on the mode of solidification.

The thermal properties of the mould material, together with those of the material to be cast and the pouring temperature, will determine the (average) cooling rate and the pattern of solidification. A change in the mould material is likely to affect the cooling rate more than it affects the pattern of solidification; consequently, a change in the distribution of temperature gradients is only likely to occur if different materials are used for various sections of the mould, i.e. the temperature gradients are made more acceptable by causing some parts of the mould to have a higher heat extraction rate than others. There is, unfortunately, a dearth of good information on this aspect of controlling solidification.

It is gratifying then that there has been more experimental work on gating and feeding methods[2,3,4] although almost all of this has shown a definite disadvantage in bottom-gating and top-feeding. This caused some surprise when the papers were first published because bottom-gated and top-fed castings were very common, but the experimental work showed, consistently, adverse temperature gradients of the order 20-100C° (68-212F°).

Side-gating (at the parting-line) and top-feeding is apparently a little better and RUDDLE[5] suggests that gating into feeders at the parting-line may be a better technique. Nevertheless considerable advantage has been found in bottom feeding for high-strength high-density aluminium alloy castings, in the Cosworth process.

The mould-reversal method[6] attempts to avoid the disadvantage of bottom-gating and top-feeding, by pouring through a bottom feeder and reversing (i.e. inverting) the mould immediately pouring is completed. In this way, gravity feeding can take place into a cavity in which a good temperature distribution has already been established.

Castings of alloys having a long freezing range are characterized by a porosity, distributed throughout the casting either in a reasonably uniform manner or with a tendency to concentrate most porosity at the heat centre.

Some light was thrown on the effects of feeding by JACKSON's[7] experiments with variable-sized, plaster-sleeved feeders. With no feeder, the porosity was about 3 per cent, distributed but with a higher density near the heat centre. Furthermore, sinks which formed at the surface accounted for most of the solidification shrinkage. As the volume of the feeder increased, the sinking appeared to diminish and the castings became sounder, i.e. the porosity decreased. Further increase in feeder volume established directional solidification, but the porosity passed a turning point and began to increase. Jackson concluded that the chief virtue of a feeder with these alloys was to impart a hydrostatic pressure to the pasty mass, thus preventing surface sinks. What is also clear from his work is that the establishment of directional solidification is not, of itself, sufficient to ensure a sound casting in a long-freezing-range alloy.

Grain size is known to have some bearing on the problem, and CIBULA[8] has found that grain-refined bronzes and gun-metals tend to show a layer porosity which is absent in their unrefined counterparts. A more definite influence is the nature of the temperature gradients. RUDDLE[9] and WALTHER et al.[10] for an aluminium-$4\frac{1}{2}$ per cent copper alloy have shown that strength and soundness are improved by higher temperature gradients.

A superficial explanation of this is that a high temperature gradient reduces the distance between the beginning and end of freeze waves, i.e. it makes an alloy appear to have a much shorter freezing-range than it has with a low temperature gradient. What is probably more to the point is that the interdendritic channels in the pasty region will be short and consequently the flow of liquid along the channels will be impeded less. It would thus appear that the initial improvement in Jackson's experiments may well have been due to the hydrostatic effect, but that the subsequent deterioration in quality is attributable to worsening temperature gradients, i.e. the volume of casting-plus-feeder is increasing at a greater rate than superficial area, thus leading to an increase in solidification time, i.e. lower temperature gradients.

The influence of the length of the interdendritic channels is illustrated by the cast iron family; the flake-graphite irons (in which the beginning-and-end of freeze fronts are closely-spaced) have short interdendritic channels whereas the long-freezing-range nodular irons have longer channels which are, furthermore, made even longer by the general expansion caused by the enlargement of the carbide grains as the graphite forms inside them. The latter type of casting thus requires very special methods to induce directional solidification, whereas flake graphite irons can be successfully cast with little or no provision for feeding.

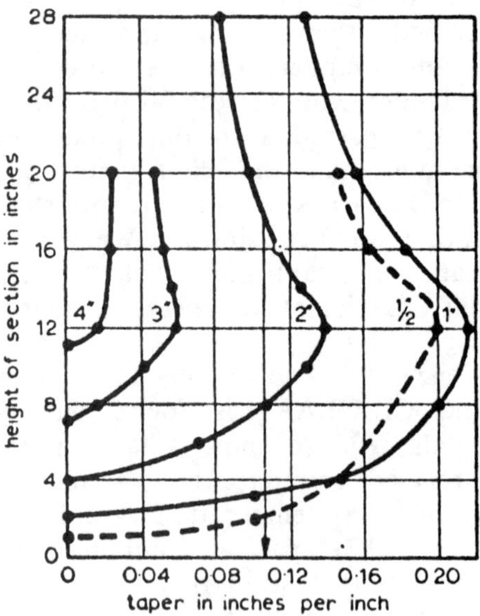

Fig. 5.1 Minimum padding (taper) for radiographic soundness
in steel plates of various heights and widths

The casting of large thin plates is particularly difficult, as exemplified by the research work of BRIGGS et al.[11], BRIGGS[12] and DUMA and BRINSON[13] which was directed towards perfecting the use of 'padding', i.e. adding metal to any thin uniform section so that it would become tapered with the largest section adjacent to the feeder. Experiments were conducted to discover the minimum amount of padding required to ensure radiographic soundness and Fig. 5.1 shows some results for various plates.

This raises the whole question of optimizing feeding arrangements and, while this may be done fairly satisfactorily for a specific casting (given the necessary time), knowledge of the effects of size and shape of feeder in general, is incomplete and, to some extent, contradictory.

Since the primary goal is to minimize heat losses from the feeder, the superficial area of the feeder should be as small as possible. A volume/area ratio V/A can be used as in Chrovinov's equation, and this is supported by measurements shown in Fig. 5.2 (after BRIGGS[12]) for steel castings weighing 14 kg.

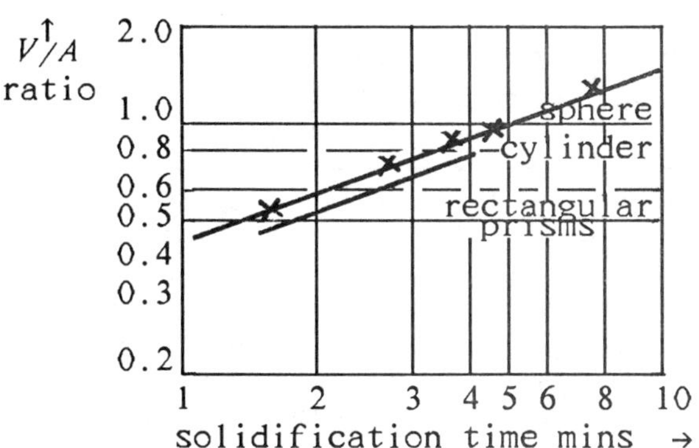

Fig. 5.2 Effect of V/A ratio on solidification time

CAINE[14] suggested the use of a relative freezing time F defined by:-

$$F = \frac{V_f/A_f}{V_c/A_c} = V_r \frac{A_c}{A_f} \qquad (5.1)$$

where f and c refer respectively to the feeder and casting. He then assumed a relationship between F and the relative volume V_r of the form:-

$$F = \frac{\alpha}{V_r - \beta} + C \qquad (5.2)$$

on the basis that when all the casting solidifies virtually

instantaneously the feeder need supply only the solidification contraction whereas if casting and feeder solidify at the same rate the feeder volume must be very large.

Using a figure of 0.05 for β, 0.12 for α and C = unity (for a sand-covered feeder), Caine arrived at the numerical expression:-

$$F = \frac{0.12}{V_r - 0.05} + 1 \qquad (5.3)$$

It is evident that better results can be obtained by reducing heat loss from the feeder. Experiments with a radiation shield or with an insulating sleeve show their advantages (see Table 5.1).

TABLE 5.1 *The effects of radiation shields and insulating sleeves in the case of 4 in $\times$ 4 in diameter cylindrical feeders*

Material	Solidification times in minutes			
	Open feeder	With radiation shield	With insulating sleeve	With radiation shield and insulating sleeve
Aluminium	12.3	14.3	31.1	45.6
Copper	8.2	14.0	15.1	45.0
Steel	5.0	13.4	7.5	43.0

It is even possible to provide heat using exothermic sleeves, but these may be wasteful of heat. A similar principle is used in the heated runners for polymer moulding.

5.2.1 *Variance in the Properties of a Casting*

It is not proposed to discuss variance in the properties of ingot castings since it will be of a lower order than that for more complex castings and much of the variance disappears as a result of subsequent metal-working operations. It should be obvious, however, that the complex physical conditions which pertain during the cooling of a cast component, must lead to local variations in the main properties of the material.

Consequently, a considerable amount of testing work is needed if a comprehensive report is to be made on a given casting. This is normally only justified in research and in components where the consequences of failure could be serious, or even fatal. Alternative procedures are the casting of a separate test-piece or modification of the component to permit a cast-on or integral test-piece (coupon†). Results from the latter type of test-piece will be more consistent with the properties of the casting, but the separate test-piece may be regarded as more indicative of the quality of the alloy. In each case, however, no regard is being paid to the known variance in properties and the results can only indicate the order of magnitude. They have, nevertheless, the advantage of economy in time and money. A complete test to destruction of one or more components is realistic, but is more an evaluation of the component, as designed, than a material testing procedure. It should be stressed that any testing of castings is not only to ensure good products, but also to control the process in the foundry.

Grain size is often considered to have an important effect on mechanical properties, but it is so difficult to isolate grain size as a single variable that conclusions which have been drawn may not always be valid. Published work shows that as the grain size of a material becomes smaller there is improvement in mechanical properties; this is generally true, although larger grain size sometimes gives higher *creep* strength.

5.2.2 *Residual Strains*

The problem of residual strains due to non-uniform shrinkage is a very real one and much depends on the geometry of the component. However, there are other, more general factors and these have been studied by FORTINO[15] with the aid of the special test-piece shown in Fig. 5.3. The upper, notched section of the ring is much heavier than the lower section and is provided with two fins or ridges. The lower section is flanged to assist in heat dissipation and resistance to bending. The test is very simple. The casting is allowed to cool and then sawn at the notch. During cooling, the slender, lower portion cools and shrinks first.

Later the upper section begins to solidify and, of course, its shrinkage is inconsistent with that of the lower section; thus a state of residual strain is set up. After the ring has been sawn through the notch, the gap is longer than would be expected from the sawing alone because of the relief of the stresses caused by the residual strain.

†Any material removed from a component for testing purposes is referred to as a *coupon*, in the American literature.

The amount of movement should be an indication of the stress level which existed in the uncut ring.

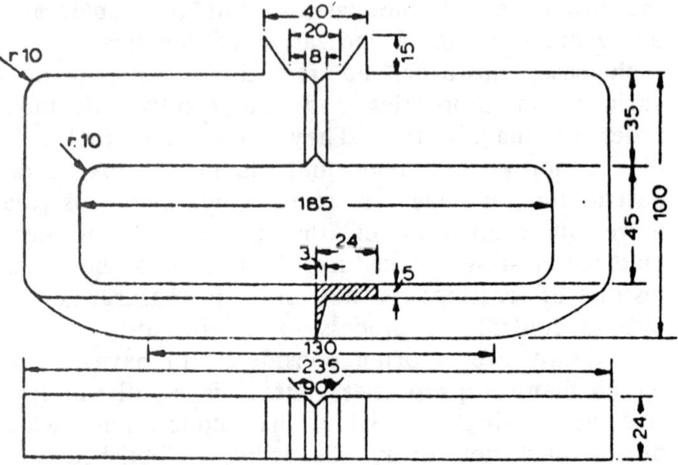

Fig. 5.3 The Fiat test-piece (after FORTINO[15],
by courtesy of *La Fonderia Italiana*)

The technique appears to be satisfactory inasmuch as repeated tests under constant, controlled conditions gave a scatter of only ± 4-5 per cent.   Three stages in cooling may be discerned in this test:-

(1)   a preliminary stage of about 15 min, during which the strain decreases quite rapidly;
(2)   a secondary stage, lasting about 30 min, in which the strain rises to a figure comparable to, or even in excess of, the maximum recorded in the preliminary stage;
(3)   a final stage, of 10-15 min, during which the general level of strain remains unchanged, but oscillations of ±10-20 per cent occur.

This pattern of events can be correlated with that of the temperatures in the upper and lower parts of the test-piece.

Using this form of test it was established that the level of residual strain rises markedly with increase in pouring temperature and falls almost linearly with increase in the carbon content of grey irons.   The work also showed that tempering and slow cooling techniques can reduce the residual strains by 75-95 per cent.

5.3 Forming from the Particle State

The various casting processes and all the processes described in the companion volume[16] use metal which, at some stage during or before the process, has been melted. This would have created increasingly difficult problems in recent years, with the gradual increase in operating temperatures, had it not been for the sintering technique.

There have been significant advances in melting and casting techniques but there are certain applications in which casting is not a possible production method; namely:-

> porous compacts, e.g. for bearings and filters;
> alloys with very high melting points, e.g. Mo, W, WC-Co;
> alloys that are mutually insoluble e.g. Pb, Cu-Zr, or with large density difference;
> alloys with poor casting qualities, e.g. Cr-Co heat-resistant alloys;
> fine dispersions e.g. WC-Co, SAP (sintered aluminium powder);
> alloys machinable only in a "green" state, e.g. gas turbine alloys;
> mixtures of metals and non-metals, e.g. cermets;
> ceramics and reaction-bonded compounds, e.g. SiN, Sialon;
> certain re-entrant and other complex shapes.

With this in mind, it is proposed to discuss the deliberate steps which are taken to confer certain properties on sintered components.

5.3.1 *Enhanced Properties at High Temperatures*

Several attempts have been made, in recent years, to use the sintering technique to enhance the high-temperature properties of materials which are conventionally regarded as already being high-temperature alloys. For example, OLIVER[17] reports increases of the order of 25 per cent in the creep strength of sintered Nimonic 100 compared with wrought Nimonic 100.

It would appear that this increase is due to included oxide or some other dispersion and the major technical problem in achieving creep strength (i.e. adequate strength over prolonged periods) is to maintain the constancy of dispersion of particle size with time.   In view of the high free energy of formation of oxides, most success might be expected with inclusions of oxide, provided they are not too large.   GREGORY and GOETZEL[18] have investigated the addition of Al_2O_3, MgO, and ThO_2 to certain Nimonics and the benefits, compared with cast and wrought Inco or Inconel are appreciable.   Creep curves given by BRUCKART et al.[19] for molybdenum, show the greatest improvement with additions of TiO_2 or ZrO_2, and a smaller improvement (although still markedly better than unalloyed molybdenum) with MgO and SiO_2.

On the debit side, the impact strength of all these included oxide materials is poor compared with conventional high temperature alloys – GLENNY and TAYLOR[20] quote a metal/TiC combination as being the best, yet still having only 10 per cent of the impact strength of cast Nimocast 258. Poor resistance to mechanical shock is, unfortunately, often accompanied by poor resistance to thermal shock.   Not a great deal is known fundamentally of the poor mechanical and thermal shock resistance of these materials but reference may be made to certain papers[21-24].

BREWER and PEARSON[21] besides giving an account of the initiation and propagation of cracks, propose an indentation method of evaluating toughness.   The whole subject is considered in greater detail elsewhere[25].

5.3.2 *Manufacture of Various Magnetic Materials*

In order to produce satisfactory strip magnetic material of high permeability, non-metallic inclusions and impurities which enter into solid solution must be eliminated as far as possible. Powder metallurgy offers greater potentialities for achieving this than casting techniques since better control can be exerted on the purity of powder metal than bulk.   JONES[26] estimated that, at - the time, most manufacturers 'have to reject or down-grade at least half of their metals'.

The subjection of (ferro-) magnetic materials to alternating-current excitation involves following the cycle of hysteresis loop at a rate of fifty times a second upwards. The loss of energy due to this is called the hysteresis loss and, within limits, is proportional to frequency.   There are further losses due to eddy-currents which are proportional to frequency squared and the failure, under certain (economic) circumstances, of the classical method of reducing these losses, viz. laminations, has led to the development of magnetic cores sintered from compacts of insulated magnetic powders.

These have the advantages of wide and accurate control over permeability and the resistance which controls the eddy-current loss, increased magnetic stability and the usual advantages of powder metallurgy with regard to sizes and shapes. The subject is very broad and, for further information, the following works may be consulted:-

the volume edited by RICHARDS and LYNCH[27] for
    soft-magnetic materials;
the review by SMIT and WIJN[28] for ferrites;
the review by SCHALLERER[29] for two-state or square-loop
    ferrites (these are used as memory devices in computers).

5.3.3 *Porous Materials*

From an engineering viewpoint, these materials may be intended to be:-

(1)   permeable to fluid flow (gas/liquid separators, aircraft-engine coolers, aerators, etc.);
(2)   capable of filtering out suspensions from gases and liquids;
(3)   capable of retaining liquids (usually lubricants) over long periods of time.

The study of flow through porous media (e.g. see SCHEIDEGGER[30]) is of concern only to the extent of needing to know the characteristics which have to be produced in the sintered component. This, together with the ease with which the material concerned sinters, will govern whether or not pressing is accomplished. If pressing is used, then it must be carefully controlled because a <u>controlled</u> density is now the goal rather than the less critical target of achieving the greatest possible density.

In this connection, use has been made of pore-forming materials such as certain metallic oxalates and acetates, which occupy space during pressing yet volatilize readily in the early stages .of sintering. It should be re-emphasized here that permeability is being sought rather than porosity (see Chapter 3).

These porous products are not cheap, because the cost of the powder in them is, itself, high and, although they may perform adequately in practice, there is little doubt that there are ways in which they could be improved, e.g. a directional, rather than random porosity, and the development during sintering, of a satisfactory strength without causing the 'throats' between particles to become so small that they restrict the flow of material in the liquid state.

The porous bearing is one of the commonest examples of this type of product. The retention of the oil (over and above the small loss resulting from the lubricating action) is of paramount importance and is subject to the surface tension of the oil, its temperature dependence, the parameters constituting the lubrication function, etc. This subject, outside the scope of powder metallurgy, yet of vital importance in manufacturing such bearings, has been studied and reported by MORGAN and CAMERON[31].

In concluding what is, of necessity, a cursory survey of a wide field, reference may be made to the uses of powder metallurgy in producing certain components with two or more metallurgical phases, e.g. electrical static contacts of copper-nickel, cadmium-silver, etc.; electrical rubbing contacts of silver-graphite, copper-graphite, etc.; non-porous bearing materials of copper-lead (or copper-tin-lead), silver-PTFE, etc.; friction materials in which a metallic matrix is used to bond non-metallic friction materials; diamond-impregnated tools. Again, the excellent review by JONES[26] is recommended for study.

5.4 Deformation Processes

5.4.1 *The 'Large-scale' Metal-working Processes*

Hot forging, hot rolling, and hot extrusion generally improve the quality of ingots. Their principal effect is to break up the cast structure of an ingot to produce smaller grain size, and to close and weld up cavities or blow-holes which are not too severe. The processes are not easy to control, as will have been realized from the discussion in Chapter 2, and any departure from a well-tried sequence of strain-rate and temperature history will usually lead to defects, especially if the metal is a complex alloy. The geometry of the tools is also important, since undesirable (tensile) stress systems may be introduced.

Considering forging first of all, the geometry of the tools which is acceptable is defined in Chapter 3 of the companion volume[16]. One of the main considerations in forming is that the metal should be worked right through to the centre of its mass, and this will not be achieved if the indenting tools or platens are too narrow. Thus it is inadvisable to forge a round bar between flat dies, since the contact area will of necessity be small. Tensile stresses are induced at the centre which may lead to the formation of cavities there. This is the principle which is utilized in the Mannesman process of tube rolling, as discussed elsewhere[16].

If the part being forged is too long and narrow, buckling may occur, leading to the formation of folds in the surface layers. Folds can also be formed under forging conditions in which the deformation is localized at the surface layers, either because the geometry of the tools is incorrect, in slow speed forging, or simply because the forging speed is very high, as in hammer forging. Another cause of folding occurs in forging between closed dies when the 'flash' formed in one operation is squeezed into the metal in a subsequent operation.

Cracking and lamination of forgings can occur if too heavy a reduction is attempted. Heavy reductions produce anisotropy in which the transverse strength and ductility are low, leading to fracture during forging. This situation often arises at the 'flash', where reduction has been very high, and 'flash cracks' are produced which form a source of potential weakness for the whole forging.

There are several causes of laps and seams or folds during forging, as well as other defects, and a useful discussion is given by WOLDMAN[32]. He points out that laps or folds formed during one stage of a forging can be opened up during subsequent operations and cause cracking of the whole piece. Cracks may also result from forging at too low a temperature, whilst too high a temperature will cause burning of the metal. 'Burning' is defined by POLUSHKIN[1] as heating the metal to a temperature so close to its melting point that permanent injury is caused to the metal by intercrystalline penetration of oxidizing gases, or by incipient melting. Burnt metal often contains voids which if oxidized or contaminated in some other way will remain. Although such voids can be closed by working, their injurious effect still remains in the metal which may appear normal on the surface — this deceptive appearance is a dangerous feature of burnt steel. Segregation of impurities is often associated with the formation of 'pipe' during casting and this may find its way into the forging if the ingot has not been properly cropped.

Forging develops metallurgical anisotropy, which is probably the most important effect on the billet. This is generally referred to as grain flow, or fibre, and the direction of the grain flow in relation to the working stresses to be imposed on the part is extremely important. For example, a gear blank machined from bar stock might have the grain structure illustrated in Fig. 5.4(a), whereas a gear blank forged by upsetting would have the grain structure shown in Fig. 5.4(b). The grain flow should be, as far as possible, in the direction of the maximum stresses. Therefore the gear teeth formed on the gear blank of Fig. 5.4(b) will be much stronger than those formed on that shown in Fig. 5.4(a). Flow lines in steel forgings can be shown up very easily by macro-etching a ground and polished section. (Immersion for 30 min at 160°F - 170°F in a solution of equal parts of hydrochloric acid and water will reveal the structure.) The fibrous structure is usually attributed to segregation of inclusions such as silicates and sulphides, and is virtually unaffected by subsequent heat treatment. It is analogous to the 'grain' of wood and produces similar anisotropy of properties.

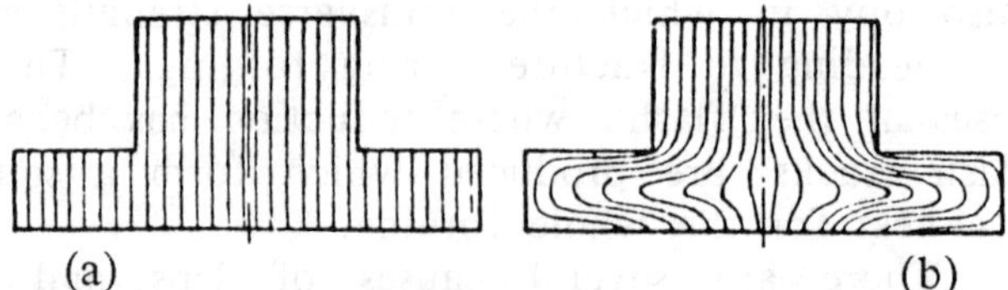

Fig. 5.4 Typical grain structure of gear blank:
(a) machined from bar stock,
(b) forged

The main distinction between hammer forging and press forging is in the resultant deformation imposed on the piece. In hammer forging the deformation is localized near the surface, where a fine-grained structure will occur during heating, whereas the material at the centre may have been deformed by only a small amount, so that the grain size may be relatively large there. Thus press forging must be used for large components if it is required to deform the metal right through the thickness. Impact loading of the body-centred cubic structure often leads to twinning on the (112) plane, and in α-iron this gives rise to characteristic surface markings known as *Neumann Bands*. This twinning provides a barrier to the movement of dislocations so that surface hardening of a part which has been hammer forged or explosively-formed can often be brought about with little or no deformation. Such effects are reduced, of course, if carried out at high temperature when 'annealing-out' of defects will occur.

Both in hot forging and in hot rolling close control of the surface must be achieved, so that scale, zones of high impurity concentration, or any other defects such as inclusions are eliminated, which would otherwise be impressed into the surface of the deformed metal. Various methods may be used to remove surface layers, one of the most popular for rolling being by *scalping* the billet. During hot forging the scale may be removed by blasting air or dry steam on to the part.

A discussion of the various defects produced by the processes under discussion could easily develop into a catalogue of procedures that have been established over the years to avoid their occurrence, such as correct forging temperatures for different metals, correct surface preparation and prior heat treatment. Although there are dangers in generalization, it is only by generalization that a reasonably over-all picture of the considerations involved can be obtained. Thus, many industrialists have adopted the terminology and ideas used by SACHS[33]. In his book he generalizes the failures which occur in direct-compression type processes such as forging, rolling, extrusion, some of which are illustrated in Fig. 5.5, and avers that they are due to 'secondary tensions'. Sach's ideas are best put over by illustrations like those shown in Fig. 5.5, the arrows illustrating the direction of the tensile stresses which are induced either by the restraint of the dies or by the geometry of the deformation.

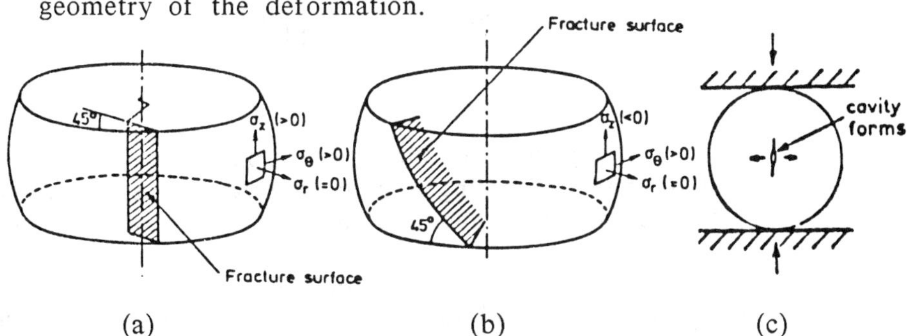

Fig. 5.5 Hot forging defects:
(a) longitudinal cracks in upset forging,
(b) shear cracks in upset forging,
(c) radial cracks in swaging

There is now evidence that many cracks follow the direction of maximum shear, as in Fig. 5.5(b). They are probably due to localized shear failure which in many instances results from a thermal shear instability. If an alloy shears slowly at a particular location and in a particular direction, the local yield strength will increase and further deformation will spread the shear zone. If, in contrast, the shear rate is very high, the heat generated will raise the temperature, causing local softening so that further deformation will be easier in the already deformed zone. These extreme conditions can be regarded respectively as isothermal and adiabatic shearing. The latter leads to an instability, limited in theory by the alloy reaching its melting point in the shear band. Whether melting actually occurs is debatable, because the shear energy close to the melting point may be too small to provide the latent heat, but there is now widespread evidence for "quasi-adiabatic" shear being the cause of many fractures in metal forming[34].

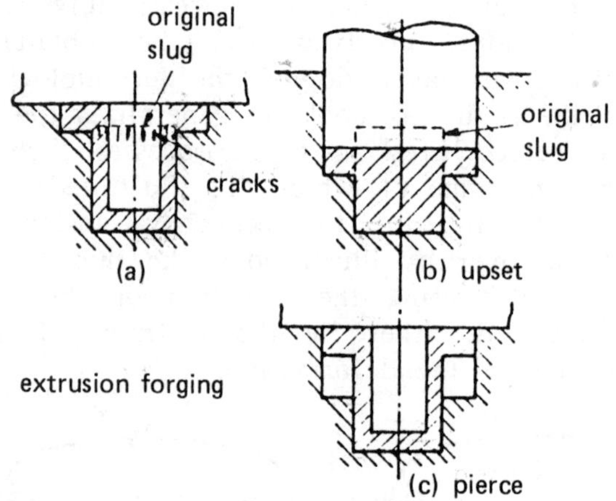

Fig. 5.6 Combined extrusion-forging:
(a) peripheral cracking due to forging in one operation,
(b) and (c) upsetting and piercing sequence to avoid cracking

Fig. 5.6(a) shows an extrusion forging which, if formed in one operation would fail by cracking at the positions shown where a great deal of stretching has to occur. If the operation is separated into two components as shown in Fig. 5.6(b) and (c) it is possible to deform the metal under substantially compressive conditions to avoid fracture.

Secondary tension in rolling may lead to the familiar *edge-cracking*, illustrated in Fig. 5.7(a), or *crocodiling* or *alligatoring*, shown in Fig. 5.7(b). Such failures are sometimes extremely difficult to avoid, and need careful control of rolling conditions. The analogous defects in drawing and extrusion are illustrated in Fig. 5.8(a) and (b), the defect in extrusion often being referred to as the *fir-tree* or *Christmas-tree* defect.

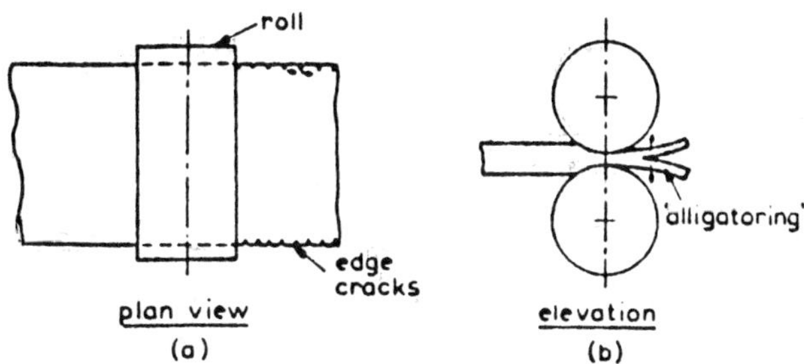

Fig. 5.7 Defects in rolling:
(a) 'edge cracking',
(b) 'alligatoring'

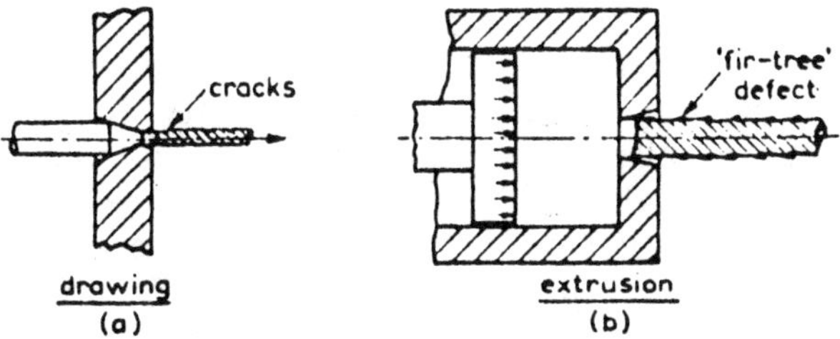

Fig. 5.8 Surface defects in drawing and extrusion

This latter defect usually occurs with the harder, more brittle alloys, as a result of trying to extrude too quickly, and the exact cause is not understood. Another defect which occurs in extrusion is that of *piping* or *extrusion defect*, when extrusion is carried on too far so that the pipe which forms in the back end of the billet is carried through into the product as illustrated in Fig. 5.9. The basic cause of this piping is that the continuation of metal flow along the die face would involve an increasing amount of redundant deformation as the ram face penetrates the plastic zone. Less energy is required to operate a mode of deformation in which shear occurs along the ram face, eventually leaving a cavity at the centre of the ram. This transition in flow mode can be retarded by increasing the friction on the ram, and conversely, very long extrusion defects may occur if the end of the billet is well lubricated. Very non-uniform flow can occur during

the extrusion of certain alloys which exhibit phase changes
near the extrusion temperature, and this will also lead to an
extrusion defect.   A detailed discussion of such defects in
extrusion is given by PEARSON[35].

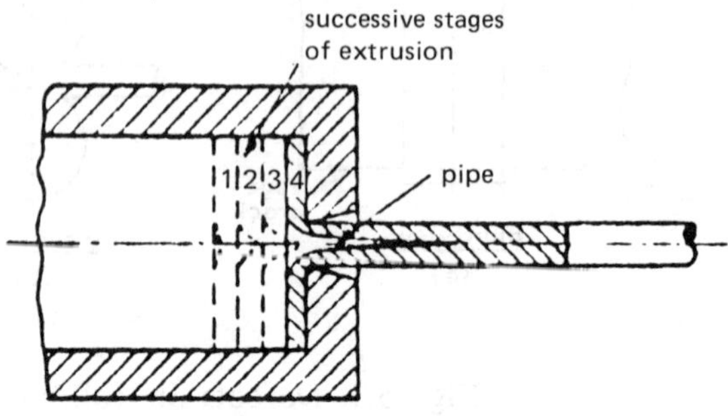

Fig. 5.9 The 'extrusion defect'

A hot-rolling process introduced a few years ago is that
of *planetary rolling,* which gives a characteristic defect on the
rolled surface in the form of a ripple due to the gap between
successive rolls passing over the contact arc.   This condition is
improved slightly if the planetary cages are rotated in the
reverse direction so that the work rolls 'climb up' the arc of
contact rather than the reverse.   Planetary mills and their
parent *Sendzimir mills* are now widely used.   As stated above,
the former give a washboard effect on the surface, which can
be rolled out by a *pendulum mill*[16].

5.4.2 *The 'Small-scale' Metal-working Processes*

Many of the *defects* which can occur in cold forging,
cold rolling, and extrusion have already been mentioned in
relation to the large-scale processes.   As a general guide it is
usually bad practice to try to form a complex shape in one
operation, as fracture inevitably occurs due to the metal trying
to flow in several directions at once.   It is much better to
carry out the simplest operations conceivable, one at a time.
This will both reduce the tendency towards cracking and also
ease the loads on the tools.   By and large, metal is more
ductile when it is hot, so that a *fracture* is more likely to
occur in the small-scale cold processes.   On the other hand,
laps and surface defects are less likely to occur, since
lubrication is more effective and scale does not form.   Indeed,
one of the more important features of cold working is the

possibility of obtaining an improved surface finish, and cold rolled sheets of steel or non-ferrous metal usually have a very good surface finish.

Perhaps the main feature of cold-working processes is the occurrence of extensive work hardening. This is generally beneficial in that the product is thereby made stronger, although ductility is reduced. Thus it is often necessary to soften the metal, either between cold-working operations, or to produce a ductile product (e.g. sheet for subsequent deep-drawing). Another important feature of cold-working processes which is often not sufficiently recognized is the possibility of introducing residual compressive stresses in the surface layers of the component, thereby enhancing fatigue strength. This is often done intentionally by *shot-peening* certain components, and by *scragging* springs, but it is not generally appreciated that the superior strength of cold-rolled threads is partly due to compressive residual stresses which have been imparted by the process. The influence of compressive mean stress on fatigue behaviour was shown in the Goodman Diagram (Chapter 1, p. 6, Fig. 1.4) and it is certainly beneficial, since compressive stresses will always reduce the likelihood of crack initiation and propagation.

Cold-forged and cold-extruded components are being used increasingly in place of parts which used to be cast, hot-worked, and machined to size, because of the more favourable economics[36]. Machining is wasteful, and it is possible to produce parts having good surface finish and dimensional accuracy by cold forging or extrusion. Also the work hardening that is imparted to the material sometimes allows the use of a lower grade alloy, the hardness being conferred by the process itself. Of course, the forces involved are high and tool wear becomes a problem with harder metals such as steel. Techniques of die design, lubrication, and improved tool materials are continually being developed to overcome these problems.

The processes of pressing and deep-drawing lead to certain characteristic defects. Wrinkling of flanges will occur unless they are properly controlled by suitably applied pressure. This wrinkling may be acceptable in certain applications, for example, in the door panels of motor cars where the wrinkles are not seen. The wrinkling is caused by the compressive stresses induced in the flange during deep-drawing, and similar buckling will occur in tubes or sections subjected to bending if the fibres under compression are not adequately supported. Cold deformation of sheet metal which has too large a grain size will give a mottled surface looking rather like, and given the characteristic name of, *orange peel*, and the Erichsen Test will reveal this defect, as already mentioned. Metals which exhibit a sharp yield point, such as mild steel, exhibit Lüders bands on deforming, giving characteristic 'stretcher strain' markings on drawn or stretched sheet. As already discussed these can be avoided by temper rolling, to introduce a slight amount of work hardening and distribute dislocations more evenly throughout the metal.

Deep-drawn components often fail due to fracture near the bottom, which occurs because the metal cannot withstand the drawing load. Bending under tension causes thinning of the sheet metal to occur in the neighbourhood of radiused tools, and fracture sometimes occurs at such points. There is little that can be done to prevent such failures, although suitable roughening of tools to prevent stretching of the metal can help. Considering the deep-drawing of a cup, it will be realized that the maximum amount of work hardening takes place in the flanges, and hardly any at all in the bottom of the cup. Thus, if a cup were required which had a fairly uniform hardness over its whole surface, it would be necessary to start with a blank which was harder in the centre than at the outside. This has led to the technique of 'differential annealing', in which a blank of, say, half-hard material is annealed by electrical induction heating which is concentrated around its periphery and softens it. When such a blank is deep-drawn the resulting part will have a more uniform hardness over its whole surface.

Another important effect of cold working, especially of sheet metal, is the occurrence of 'spring-back'. This is the manifestation of the elastic recovery of fibres which have been subjected to plastic flow, and is the reason for the occurrence of residual stresses, which may be either tensile (e.g. secondary tensions) or compressive (e.g. as produced by peening). Thus the tools required to form a component of a certain shape will rarely have the same shape as the component, and they have to be developed by trial and error. Spring-back can be reduced by superimposing a high mean stress on the bending system of stress concerned, which is an important advantage of the stretch-forming process. Smaller spring-back is also observed in explosively formed sheet components, but it is not known whether this is due to a superimposed tension or compression or some other cause.

5.5 Metal-removal Processes

5.5.1 *General Influence of Machining*

Various machining processes such as lathe turning, milling and drilling are widely used for the final shaping of metals. These are slow in comparison with forging or extrusion but are extremely versatile. Machining is wasteful of material and expensive for very large batch sizes but is unmatched for single items or small batch production. The latter is becoming increasingly popular with numerical control (NC) or computer numerical control (CNC). It also integrates well into the philosophy of flexible manufacturing systems (FMS).

Whereas the properties of products can sometimes be enhanced by forming, for example in producing stronger gears through work hardening of the blank and improved directional properties, the effects of machining tend to be detrimental, even if only mildly so.

The major influence is on the surface topography. In lathe turning, for example, there is a natural waviness on the surface due to the feed marks replicating the tool nose. Usually, unless extreme precautions are taken there will also be a superimposed roughness of much smaller wavelength, due to a tearing action and rubbing on the flank or clearance faces of the tool.

This rubbing also introduces high local stresses which may cause plastic deformation of the workpiece surface, leading to heating and to residual stresses. The major shearing process involved in chip formation and the rubbing of the new chip against the rake face of a tool will also generate heat. Much of this is carried away in the chip or swarf which often becomes red-hot, and the intense heating at the rake face may cause tool failure, but this heat has little or no direct influence on the workpiece surfaces.

A secondary consequence of the heating is the production of phase changes in steels and other alloys, which may lead to residual stresses. In extreme instances these in turn may seriously reduce the fatigue life of a product subjected to cyclic stressing. The roughness of the surface may also play a part in initiating the cracks that lead to fatigue failure. Recently it has been recognized that high-stress low-cycle fatigue is much more important than was previously believed. Critical aerospace alloys, for example, may fail after only a few thousand cycles although the fatigue life predicted from laboratory tests on carefully polished specimens may be very much greater.

5.5.2 *Temperatures in Metal Cutting*

Because temperatures can seriously affect tool life and can also induce unwanted metallurgical changes in the workpiece, it is important to consider the heat generation and flow in some detail.

The classical approach to this problem is illustrated in Fig. 5.11.

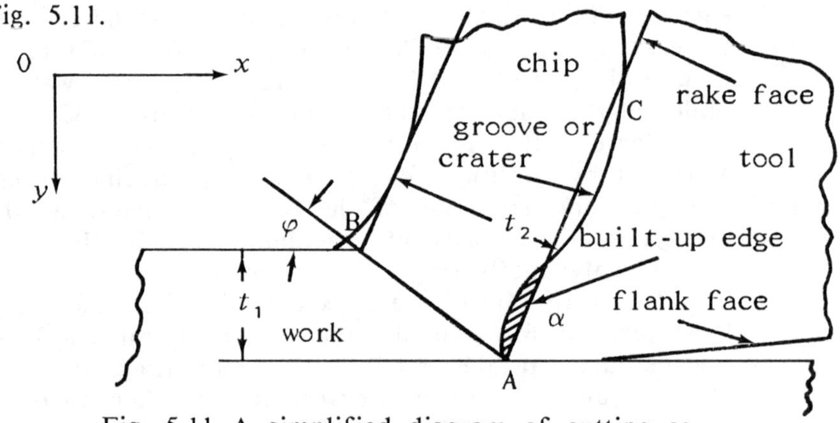

Fig. 5.11 A simplified diagram of cutting as
used for analysis of plane-strain machining

The fundamental deformation process involved in lathe cutting, planing or milling is an intense shear along a narrow zone reaching from the tool tip to the free surface. The process is usually idealized by considering a two-dimensional system, as in planing, with the tool edge orthogonal to the cutting direction and no metal flow perpendicular to the plane of the diagram.

A further simplification is obtained by assuming that the primary chip formation process is shear on a plane inclined at an angle φ to the direction of cutting. The metal thus approaches the tool horizontally in Fig. 5.11, shears along the plane AB and then flows parallel to the rake face. Since there is very restricted access of lubricant to the interface AC between the chip and the tool, and the metal surface is freshly generated, the friction there is very high, sufficient to cause plastic deformation and the formation of a secondary shear zone. The evidence for such deformation is often clear from a metallurgical section of a steel chip. Intense heat may be generated by the secondary deformation, leading to a very high temperature at the rake face.

In real cutting the tool does not stay sharp, as Fig. 5.11 implies, but wears away to form a wear land on the flank face. This also causes severe rubbing and a further heat source is created.

Temperature has proved a major limitation in metal cutting. The early carbon steel tools could be used only at very modest speeds, and a great advance was made by the introduction of high-speed steel (HSS) tools which could cut at much greater speeds and were even capable of functioning when red hot. A further advance was made by the introduction of cemented carbide tools, initially WC-Co, which is still the most widely used carbide. More recently, speciality tools with polycrystalline diamond (PCD), TiN or CBN (cubic boron nitride) cutting surfaces have been introduced with enhanced wear resistance. These are discussed by ALEXANDER, BREWER and ROWE[16] in the companion volume. Attention will now be given to the analysis of heat flows and temperature in simple plane-strain orthogonal cutting.

The classical analysis is based on the basic differential equation for heat flow in two dimensions with a moving heat source (CARSLAW & JAEGER[37]), which is:-

$$\frac{\partial^2 \theta}{\partial x^2} + \frac{\partial^2 \theta}{\partial y^2} - \frac{V}{\kappa}\frac{\partial \theta}{\partial x} = \frac{1}{\kappa}\frac{\partial \theta}{\partial t} \qquad (5.4)$$

where κ the thermal diffusivity and V the velocity. θ is the temperature at a general point whose coordinates are (x,y).

For simplicity it is assumed that heat is generated uniformly along the shear plane. A small fraction of the deformation energy is stored as elastic strain energy but this can be neglected in comparison with the work of plastic deformation so, to a good approximation, the total heat Q_T generated during cutting is given by:-

$$Q_T = \frac{W}{J} = \frac{FV}{J} \qquad (5.5)$$

where F is the cutting force, V the cutting velocity and J the mechanical equivalent of heat. The heat generated along the flank face is neglected in a simple analysis for a sharp tool so:-

$$Q_T = Q_S + Q_R \qquad (5.6)$$

where S and R relate to the shear plane and the rake face respectively. These quantities Q_S and Q_R can be distinguished experimentally because:-

$$Q_R = \frac{F_R V_R}{J} \qquad (5.7)$$

and F_R the force component parallel to the rake face can be measured with a suitable tool dynamometer and V_R is the velocity of the chip along the tool face.

In continuous cutting, the pattern of temperature distribution settles quickly to a steady level, so the time-dependent element in Equation 5.4 can be disregarded. In addition the conduction of heat in the direction of motion (along Ox) is small in comparison with the transport of heat by the moving metal itself. Heat conduction in the direction Ox is also negligible in comparison with conduction into the workpiece, or more correctly into the "undeformed chip" material below the shear plane AB, in the direction Oy.

Equation 5.4 thus reduces to:-

$$\frac{\partial^2 \theta}{\partial y^2} - \frac{V}{\kappa} \frac{\partial \theta}{\partial x} = 0 \qquad (5.8)$$

The parameter Vt_1/κ is known as the *Thermal Number*, usually expressed in metal cutting as:-

$$R_T = \frac{\rho c V t_1}{k} \qquad (5.9)$$

where ρ is the density, c the specific heat, V the cutting speed, t_1 the undeformed chip thickness (Fig. 5.11) and k the conductivity. Thus:-

$$\frac{\partial^2 \theta}{\partial y^2} - \frac{R_T}{t_1} \frac{\partial \theta}{\partial x} = 0 \qquad (5.10)$$

The first important analyses of these temperatures in the workpiece and shear zone were conducted by WEINER[38] and by RAPIER[39]. Weiner obtained a solution for the temperature distribution along the shear plane in the form:-

$$\theta \sin \varphi = (1+2X)\operatorname{erf}\sqrt{X} + \frac{2}{\sqrt{\pi}}\sqrt{X}e^{-X} - 2X \qquad (5.11)$$

which is in terms of X, itself dependent upon the position and other parameters.

Rapier found the maximum temperature θ_m in relation to the mean temperature $\bar{\theta}$ to be given by:-

$$\bar{\theta} = \frac{Q_R}{\rho c V t_1 w} \qquad (5.12)$$

Then:-

$$\frac{\theta_m}{\theta} = 1.13 \sqrt{\left[\frac{R_T}{\alpha}\right]} \qquad (5.13)$$

where α is related to the length of the shear plane.

This overestimates θ_m, probably because the actual heat source is not localized to a single line.

A revised model assumed a second uniformly distributed source along the rake face, which gave more realistic results. This is discussed more fully in the book by BOOTHROYD[40].

An iterative solution gives temperature contours in the workpieces, the chip and the tool as shown in Fig. 5.12.

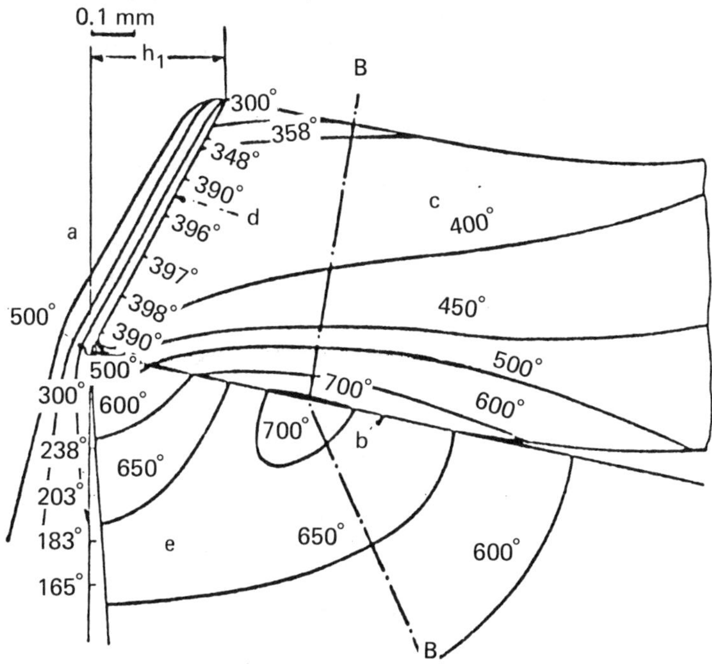

Fig. 5.12 Representative temperature contours

More recently, analysis by the Finite Element Method has been applied by SHIRAKASHI and USUI[41] and others[42]. Details of FEM are given in ALEXANDER, BREWER and ROWE[16]. This does not require the restrictions of the solution of differential equations being based primarily on equilibrium and continuity equations.

It appears possible that FEM will supplant earlier analytical methods in some metal working processes, particularly those in which the deformation is particularly severe, as in extrusion for example. Processes like strip or slab rolling are adequately solved by using traditional "slab" methods of solution, which can now be readily adapted to more accurate numerical solution by using computer methods as described in the companion volume[16] (Volume 2). In view of the additional difficulties involved in the solution of metal cutting problems, e.g. the separation of the tool tip and the complex frictional conditions, it can be expected that further theoretical development will use FEM or similar numerical techniques.

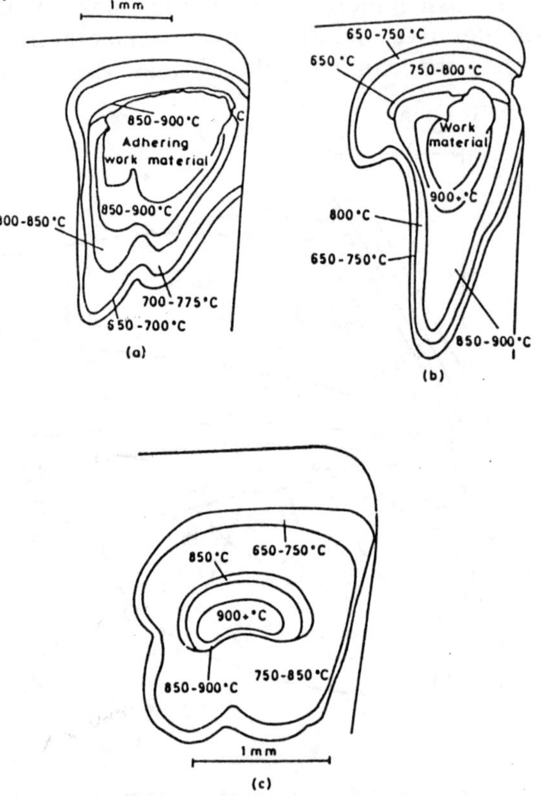

Fig. 5.13 Temperature contours deduced from
tempering structures (TRENT[43])

A quite different approach to the question of temperature has been taken by TRENT[43] in relation to tool wear. This is discussed more fully in Volume 2. Trent has prepared many metallurgical sections of HSS cutting tools after use[44]. By a special technique based on the observation of structures produced by a range of tempering times and temperatures he was able to define temperature contours. A typical set of results is shown in Fig. 5.13. This includes the effect of the contact on a worn flank and agrees well with external measurements and also supports in general terms the numerical analysis by CHAO and TRIGGER[45].

Particular features to observe in considering the influence of machining on properties are the generation of heat on the flank face, which becomes much worse as wear proceeds, and the penetration of heat into the workpiece from this source and also, to a lesser extent, from the primary shear zone. In most industrial cutting of steels using a good lubricant, the temperatures at the workpiece surface do not reach any transition temperatures, so little irreversible damage is done.

In grinding, however, there is a much greater possibility of transformations being induced, leading to local volumetric changes that can produce severe residual stresses. When considering this problem, the action of a grinding grit can be considered to be that of a tool having a highly negative rake angle with a large wear flat. It is generally accepted, as discussed more fully in Volume 2, that the cutting force increases as the rake angle α becomes less positive, and that the increase is rapid for large negative rake angles. The shear plane (as idealized) descends further into the undeformed chip, so for both reasons more heat diffuses downwards into the workpiece, in addition to that generated on the worn flank.

As the speeds of grinding are also very high, typically 300 m/min, very high temperatures are generated. In extreme conditions these lead to visible "metallurgical burn", which is a brown oxidation mark on the workpiece, or even to surface melting. It is usual to take care to avoid these obvious features but conventional grinding, and especially grinding with a hand-held machine, can easily produce temperatures sufficient to cause phase changes in the workpiece surface. Cooling then produces high residual tensile stresses, sometimes approaching the annealed yield stress of the alloy. These can seriously reduce the fatigue life and stress-corrosion resistance of the product, and major fractures have been attributed to this cause. Fig. 5.14 shows some laboratory results.

The plastic flow at the product surface produces compressive residual stresses that can in principle increase the fatigue resistance, but these are usually very small and have little effect. As can be seen from Fig. 5.14, the depth of penetration of the residual stresses is small, usually not exceeding 0.1 mm. It is in fact possible to remove metal rapidly in the initial stage of grinding and then to revert to a much slower "gentle" grinding to remove the heat-affected zone, but this requires careful control (MITTAL and ROWE[47]).

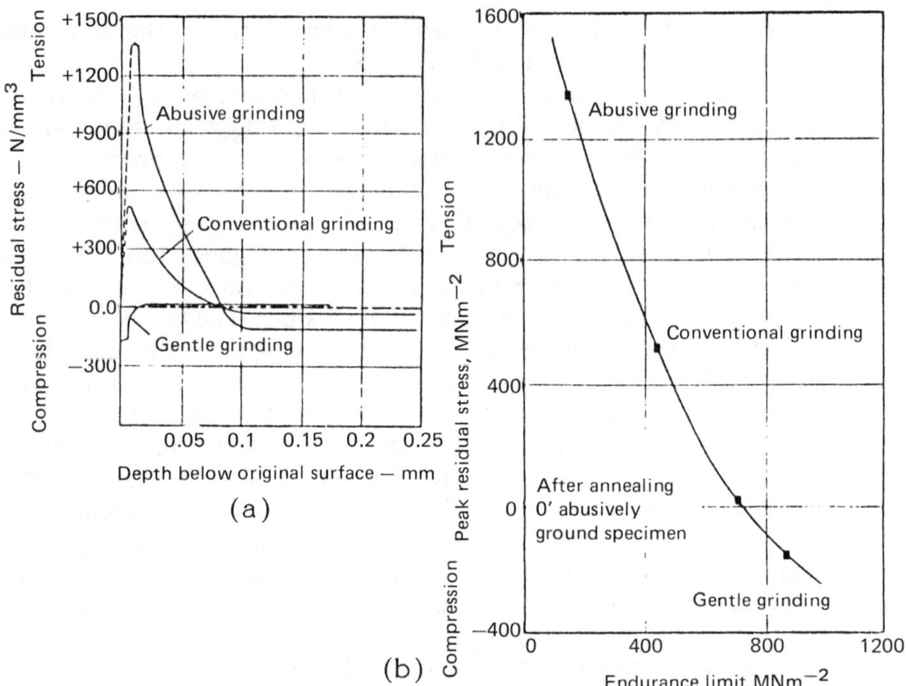

Fig. 5.14 Residual stresses (a) and consequent
fatigue life (b) of ground steel components[46]

The common practice of "sparking out" on a surface
grinder, by allowing grinding to continue after the infeed has
been disengaged, helps to some extent in this respect. The
elastic recovery of the wheel assembly and the general
machine structure reduces the radial force as sparking out
proceeds but this is not, of course, under independent control.

High residual stresses may also be induced by heating in
nominally stress-less processes such as EDM (electro-discharge
machining[48]) or laser drilling[49].

5.5.3 *Mechanical and Topographic Features*

Much attention has been given elsewhere to the forces
involved in machining, particularly to their analysis and
explanation. In terms of the effect of machining on product
properties these play a secondary role though they may set
limits to the rate of removal or the accuracy attainable.

A more important feature for most uses is the surface
roughness produced on the workpiece. There is a theoretical
limit to the smoothness that can be produced with a given

tool at a fixed feed rate, although super finishes of mirror quality are possible by special "nanometer range" machining techniques of diamond turning[50].

Fig. 5.15 illustrates this limit for a sharp angular tool (a) and for a tool with large radius (b).

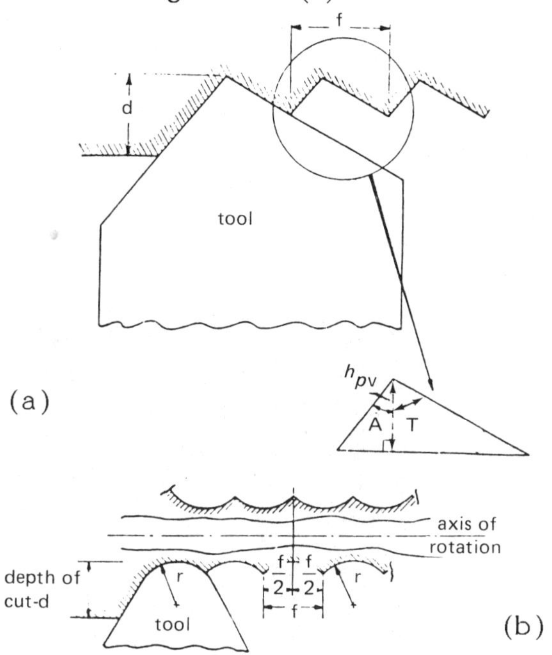

Fig. 5.15 The theoretical surface roughness produced by (a) a sharp-angled tool and (b) a large-radius tool nose.

Surface roughness is now measured as Roughness Average R_a, according to S.I. practice, though Centre-Line Average (CLA) and Arithmetic Average (AA) persist in the English-speaking world, with *Rautiefe* in German and Russian literature. There are many other measures of the height distribution, frequency, and peak radius etc.

R_a is defined as the mean departure from the mean centre line, as follows:-

$$R_a = \frac{1}{L_0 - L_1} \int_{L_0}^{L_1} |y - \bar{y}| \, dx; \quad \bar{y} = \frac{1}{L_1 - L_0} \int_{L_0}^{L_1} y \, dx \qquad (5.14)$$

For a single triangular profile, as in Fig. 5.15(a), the R_a value (or CLA or AA value) is $\frac{1}{4}$ of the peak to valley height h_{pv}. In terms of the approach and trailing angles A and T,

$$R_a = \tfrac{1}{4} h_{pv} = \tfrac{1}{4} f/(\tan A + \tan T) \qquad (5.15)$$

In addition to this "ideal" profile or *waviness*, there will usually be a finer-scale *roughness* associated with tearing of the metal at or near the workpiece surface. With a sharp tool of large positive rake angle and good lubrication the surface may be clean and reflective (CHILDS and ROWE)[51] but such tools cannot be used or will not last long enough when cutting steels.   The usual 0 to +6 degree rake angle leaves a poor surface at low speeds, aggravated by flank land rubbing.   The surface deteriorates even further at intermediate speeds because of the influence of *built-up edge* (b.u.e.) formation, which in effect produces a rough high-friction tool nose as shown in Fig. 5.16, especially when the b.u.e. extends right round the tool nose[43].

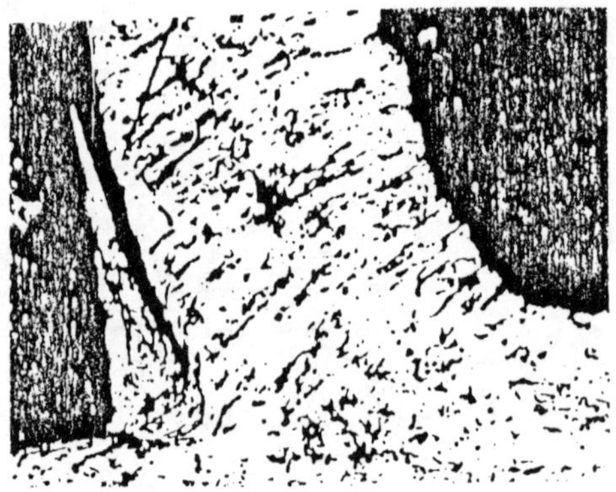

Fig. 5.16 Built-up edge formation

These influences are reflected in the variation of finish with speed shown in Fig. 5.17 for typical conditions.
Some representative values for different processes are given in Table 5.2

TABLE 5.2 *The ranges of surface roughness produced by various finishing processes*

Process	$R_a\ (\mu m)$
Lathe turning	0.3-2.5
Milling	0.5-2.5
Diamond tool finishing	0.1-1.0
Grinding	0.1-1.0
Lapping	0.04-0.15
Polishing	0.03-0.1

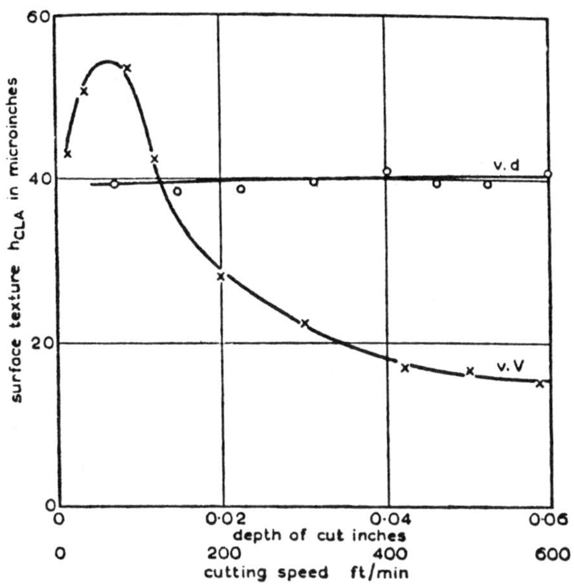

Fig. 5.17 The effect of cutting speed
and depth of cut on surface finish

5.6 Fabrication Processes

As in Chapter 4, it is proposed to exclude cold pressure
welding from the present section. It is a welding process
inasmuch as separate pieces of material are being joined and
the compressive stress level at the joining must be sufficient
to cause adhesion, but the effects of this stress system and the
plastic flow are much more closely related to the matter
discussed in Section 5.4.
Since the heat input to the workpieces and weld is
responsible for most of the physical and metallurgical changes
which are encountered in the fusion welding process, it seems
apposite to begin a discussion of what the process does to the
material by a consideration of the thermal problem.
As in machining, there is relative movement between the
heat source and workpiece but, after a short time, a

quasi-static state is reached and if it were practicable to coat the work surface with colour-sensitive paint, the colour bands after welding would all be parallel to the direction of relative movement.   (In practice, this phenomenon can often be seen from the oxidation colours.)

In formulating a theoretical approach, the usual axiom can be accepted that mathematical simplicity is sacrificed the more nearly the mathematical model represents practical conditions. Those assumptions are first made which most simplify the mathematics without causing the model to depart appreciably from actuality; for example, it can be assumed that the heat-source is located at a point and that heat losses from the surface are negligible compared with heat flow inside the workpiece.   The former assumption is reasonable in electric-arc welding, but less accurate in gas-welding.   The latter assumption is valid for most metals because thermal conductivity is usually large compared with the factors controlling heat transmission through the metal/air interface.

Secondly, it is acceptable to regard as constant those parameters which appear in the fundamental equation as coefficients of the derivatives. To all intents and purposes, this makes the difference between obtaining an analytical solution and failing completely because of mathematical difficulties. Consequently, if the temperature-dependence of, say, thermal diffusivity must be included in the problem, it is virtually essential to accept numerical methods, i.e. to be able to give a numerical answer to a specific problem, and a change in any parameter or dimension necessitates reworking the calculations[42].

For a quasi-static, three-dimensional case, Equation 5.4 is required in its full form, i.e.

$$\frac{\partial^2 \theta}{\partial x^2} + \frac{\partial^2 \theta}{\partial y^2} + \frac{\partial^2 \theta}{\partial z^2} - \frac{V}{\kappa}\frac{\partial \theta}{\partial x} = \frac{1}{\kappa}\frac{\partial \theta}{\partial t} \qquad (5.16)$$

V in this case being the speed of welding.   The following axes are assumed:-

x - in the direction of relative movement;
y - in the plane of the plate and perpendicular to the x-axis.
z - in the direction perpendicular to the plane of the plate.

The boundary conditions must now be considered and there are three approximations or tendencies for a plate of semi-infinite thickness:-

(1) imagine a hemisphere of radius $R = \sqrt{(x^2+y^2+z^2)}$, with its centre located at the point heat sources. Then, for small R, the heat flow through the surface area will tend to a limiting value Q which is equal to the total heat delivered to the plates, i.e. mathematically:-

$$\lim_{R\to\infty} \left| 2\pi R^2 . k \, \frac{\partial\theta}{\partial R} \right| = Q \qquad (5.17)$$

(2) since there is assumed to be no heat transmission from the surface to the surrounding atmosphere:-

$$\frac{\partial\theta}{\partial z} = 0 \quad \text{for } z = 0 \text{ where } R \neq 0 \qquad (5.18)$$

(3) it may reasonably be assumed that the temperature remains at ambient (θ_a) for points far removed from the source, i.e.:-

$$\theta = \theta_a \text{ for large } R \qquad (5.19)$$

The solution to Equation 5.16 with these boundary conditions is simplified by a transformation of co-ordinates but, even then, it involves a Bessel function of the first kind and zero order. A proof in broad outline has been given in ROSENTHAL[52] and will be used here, viz.:-

$$\theta - \theta_a = \frac{Q}{2\pi kR} \exp\left[- \frac{V(x-Vt)}{2\kappa} \right] \exp\left[- \frac{VR}{2\kappa} \right] \qquad (5.20)$$

It will be noted that this solution gives an infinite temperature for $R=0$, but this need be of no concern since the solution is inapplicable in any case to the volume where the metal has melted, i.e. for very small values of R. Hence the solution is valid if applied to material in the solid state.

If the plate is of finite thickness, a further term (consisting of a rapidly converging series) has to be added to Equation 5.20. The effect of this is to spread the heat over a wider area, i.e. the isothermal for a given temperature encloses a larger area as the thickness of the specimen is reduced. Another way of expressing this is to say that temperature gradients decrease as the plate thickness decreases.

The nature of the temperature distribution around the heat source, for a plate 0.2 in. thick, is shown in Fig. 5.18. It will be seen that temperature gradients are much steeper in front of the source than behind it. The dotted curve is the locus of points having the maximum temperature *at a given instant of time,* i.e. it is the curve that separates the material with rising temperature from the material whose temperature is falling; it will be noted that the finite time taken for heat flow leads to this curve being markedly curved.

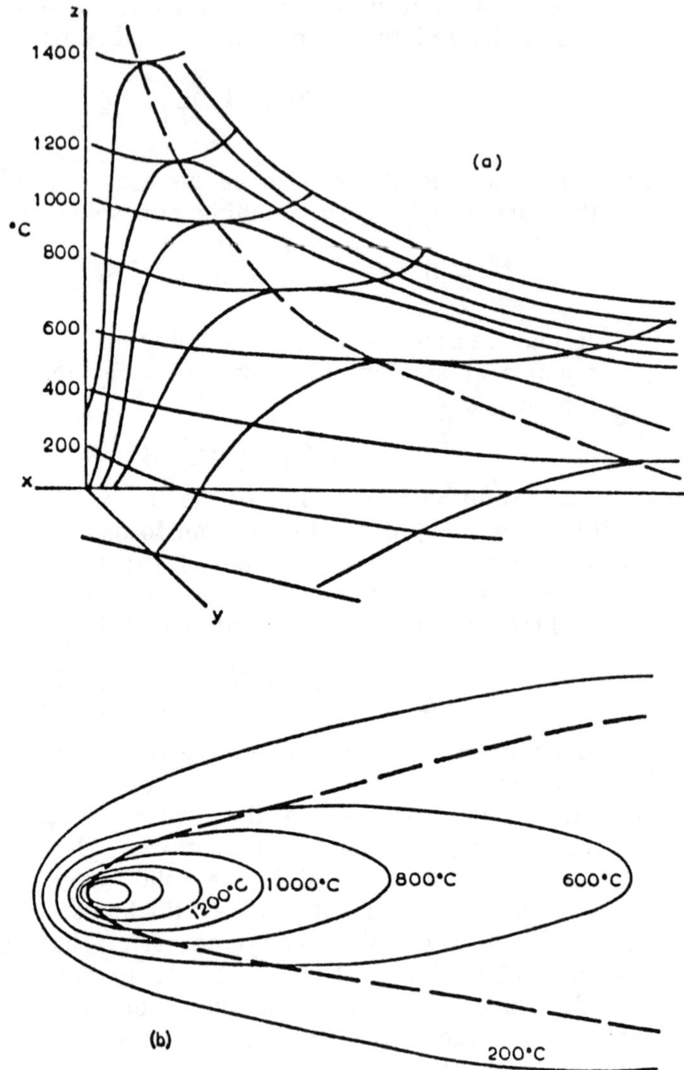

Fig. 5.18 Temperature distribution in fusion welding:
(a) 3D plot (b) isothermals in 2D plot (after ROSENTHAL[52],
by courtesy of the *The Welding Journal*)

As would be expected, the dimensions in the plane of the plate have little influence (within wide limits) but the shape of the isothermals is changed considerably by (a) the work material (the higher its thermal diffusivity, the more circular the isothermals), or (b) the welding speed (the higher the speed, the more elongated the isothermals). Preheating and the intensity of heat during welding only change the relative spacing of the isothermals, leaving their general shape much the same.

The isothermals shown in Fig. 5.18 represent a temperature distribution in space, but since the welding speed is constant, certain broad deductions about time rates of change may also be made. It will be obvious, for example, that the highest (time) rate of change of temperature is at the weld-pool boundary and that there will be a falling-off in the rate of change as the distance from the weld-pool increases.

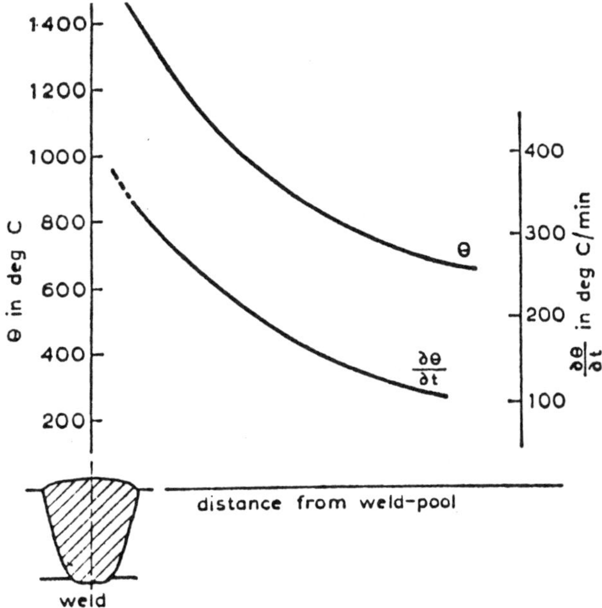

Fig. 5.19 Variation of temperature and cooling
rate with distance from the weld-pool

Hence the picture shown in Fig. 5.19 is that required for a metallurgical consideration of the effects of welding, viz. a knowledge of the temperature and rate of change of the temperature in the direction of welding. Fig. 5.19 is representative of a low carbon steel in which (like all carbon steels) the formation of martensite must be avoided. The critical cooling rate to inhibit martensite formation in such a steel can be established from known metallurgical data, thus permitting the diagram to be interpreted in terms of the likely effectiveness of the weld.   HANEMANN and SCHRODER[53], in particular, have studied the effects of rate of cooling on the formation of the Widmanstätten structure, which is likely to occur in oxy-acetylene welding and leads to brittleness.

In some cases it is possible to convert the rate of change of hardness equation from a temperature-dependent relationship to a time-dependent relationship.   The limitation is really associated with the extent to which the Arrhenius-type law:-

$$\frac{\partial H}{\partial t} = B \, \exp\left[-\frac{E}{RT}\right] \qquad (5.21)$$

can be applied to actual metals or alloys.   In this equation, H = hardness, t = time, B is a constant or $\varphi(H)$, and the terms in the exponential have their usual meaning.

Work at M.I.T. has shown that certain aluminium alloys behave according to Equation 5.21 and, by using a θ/t relationship, that equation can be put in the form:-

$$\frac{\partial H}{\partial t} = C \, \exp\left[\frac{E}{RT}\right] \qquad (5.22)$$

Microstructures predicted by the application of this equation and the relevant metallurgical diagrams agree surprisingly well with experimental observations.

The difficulties of this approach, as a practical tool, cannot be denied, but as a scientific explanation of known facts it is a significant step forward.

Furthermore, from a graph of mean grain size versus temperature, some idea can be gained of the metallurgical structure in the parent metal adjacent to the weld. Qualitatively, a very coarse grain structure would be expected near the weld; this is borne out by photomicrographs of polished and etched specimens in the case of gas welding, but in arc-welded specimens there is frequently a fairly thin layer of small-grained metal adjacent to the weld metal.   Beyond this layer there is the normal region of grain growth, but it is smaller in extent than that obtaining in a gas welded specimen.   Also, in arc welding, there is a great likelihood that a multi-pass weld will be made, in which case, all passes except the last will be normalized when the following pass is made; this does much to inhibit the dendritic structure which is so prevalent in single-pass welds and leads to a plane of weakness similar to that occurring at a re-entrant corner in casting (see Fig. 5.20).

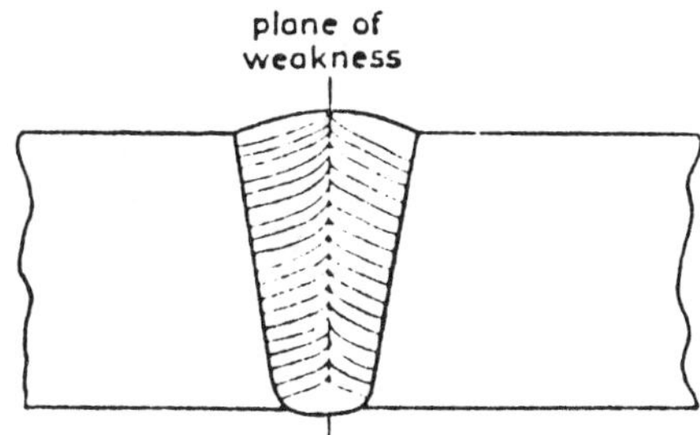

Fig. 5.20 Illustrating the plane of
weakness in single-pass welds

Chapter 2 has referred to brittle fracture and the concept of a transition temperature; although cracks can be initiated and will propagate in many circumstances unconnected with welding, it is evidently true to say that welds are potential sources of crack initiation. Furthermore, as pointed out by TIPPER[54], in her admirable survey, the solid continuum produced by welding encourages the propagation of cracks, whereas the physical breaks in a riveted or bolted structure limit the extent to which a crack can be propagated. Thus, welding and brittle fracture have been very closely associated with each other. Several of the tests which are made to assess notch-brittleness have been discussed in Chapter 4.

Consider the causes of cracking. Among the primary causes are bad welding conditions, including those which encourage the formation of nitrides, oxides, and sulphides, but it is more profitable to consider the possible metallurgical reasons.

The fact that the process of solidification is accompanied by a decrease in volume is an obvious potential source of cracks, the material which solidifies last being the most prone to this defect. This material is the weakest link, if the material as a whole is unable to withstand the stress level commensurate with the degree of 'prevented' contraction. There are numerous and complex metallurgical transformations during cooling and these may include brittle phases which are unable to withstand the stresses referred to in the previous sentence. Also, cracking may occur at either super- or sub-solidus temperatures, and the mechanism of cracking will be different in these two cases. If liquid metal is present in interstices, gases which lead to embrittlement, such as hydrogen thrown out of solution during solidification, will diffuse out more easily than if the whole mass were solid.

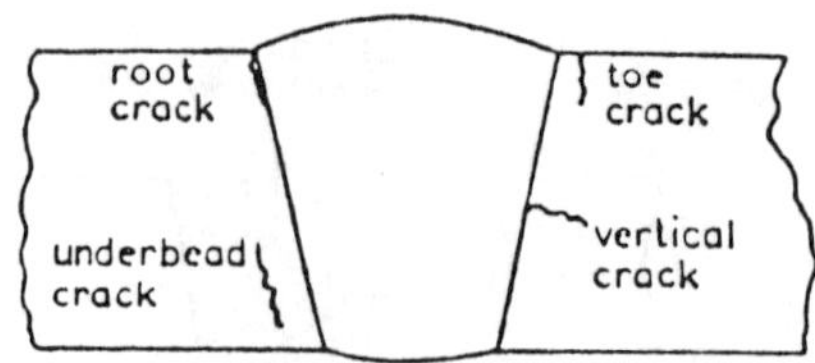

Fig. 5.21 Weaknesses in fusion welding

The chief cracks which occur in the parent metal are shown in Fig. 5.21 and are due to the internal stress system developed in a welded structure or to the effects of included hydrogen. Susceptibility to cracking is also related to chemical composition, e.g. in a free-cutting steel, it is increased by raising the sulphur content or reducing the manganese content. The sulphur exists as iron sulphide which forms a eutectic with iron at 985°C (1805°F), the eutectic composition being concentrated along the grain boundaries and leading to local brittleness; the effect of the manganese is due to its greater affinity for sulphur than that of iron, together with the fact that manganese sulphide is non-embrittling —— certainly not to the same extent as the iron sulphide – iron eutectic. Unfortunately, not all the manganese is converted to its sulphide, owing to the rapid thermal cycle of welding.

The general form of residual stress distribution is shown in Fig. 5.22. It will be seen that there is a region in which the tensile stress is quite considerable, but it is confined to a width approximately equal to twice the thickness of the plate. In a workpiece which is at room temperature before welding is commenced, hot tearing and cold cracking can only be prevented by choosing a material in which the elongation and tensile strength do not drop markedly at elevated temperature. It is not always easy to let the production process be the sole criterion in selecting a material, so where the design demands an austenitic steel, the tendency to hot tearing (often associated with columnar structure) must be accepted. Similarly, a tendency to cold cracking in low carbon and low alloy steels is a hazard which must be accepted in materials which exhibit a rapid drop in tensile strength at a fairly low temperature.

One way in which these difficulties may be resolved is to pre-heat the material but, apart from technical and economic considerations, this is limited to about 400°C because of the effect of the heat on the operator who has to weld the work. Preheating minimizes the effects of change in microstructure (including residual stresses) and permits a much higher rate of diffusion of those gases, particularly hydrogen, which cause cracking.

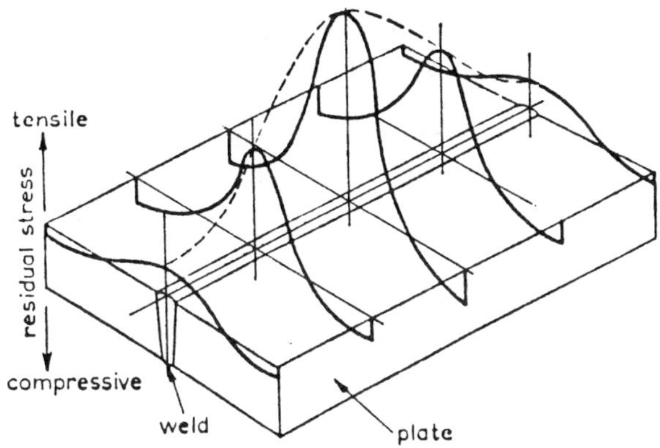

Fig. 5.22 Schematic distribution of residual
stresses in fusion welding

According to VAUGHAN and MORTON[55], crack sensitivity is a function not only of the rate of cooling, but also the speed with which the temperature drops from 150°-120°C (300°-250°F). This is consistent with the work of FLANIGAN and MICLEU[56] which showed that most cracking occurred in roughly this temperature range. This suggests that the preheating temperature plays its most important role when it governs the rate of cooling in the range of 120°-150°C; work by COTTRELL[57] indicates that this rate should not exceed about 10 C° per second but much more research over a wide range of steels is required before more than tentative conclusions can be drawn.

The effects of preheating are influenced by two other factors (thermal energy of the arc and plate thickness) to an extent which cannot be neglected. The thermal energy of the arc is most easily expressed in terms of electrical parameters, viz.:-

$$E = VIt \text{ watt sec} \tag{5.23}$$

where V, I are the arc voltage and current and t is the welding time in seconds. We may replace t by l/v where l is the length of the weld and v the welding speed, so that:-

$$E = \frac{VIl}{v} \text{ J}$$

$$= \frac{VI}{v} \text{ J/unit length of weld} \qquad (5.24)$$

The joint effect of preheating temperature and arc energy on rate of cooling is shown in Fig. 5.23, which has been plotted on log-log co-ordinates from data quoted by SÉFÉRIAN[58]. The data pertain to cooling rates at a temperature of 540°C and are for plate of 12.5 mm thickness. The dotted line is for a 25 mm thick plate, preheated to 100°C.

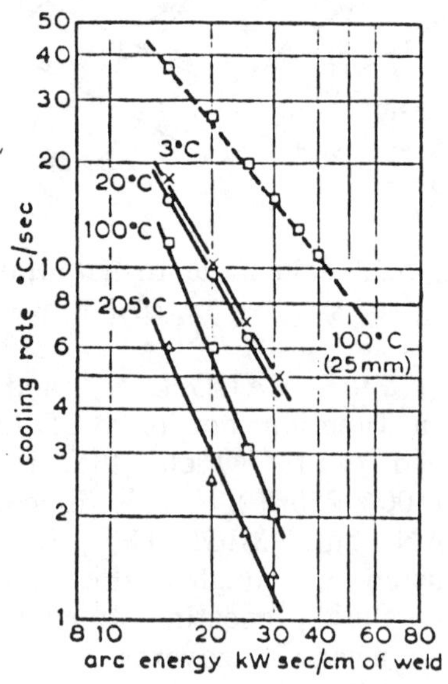

Fig. 5.23 Effect of arc energy and preheating
temperature on cooling rate

Comparison of the 100°C lines for the two thicknesses shows how much the arc energy must be increased in order to maintain the same cooling rate in the thicker specimen.

The main problem in practice, however, is the ubiquitous one of enabling the user to determine the preheating temperature validly and quickly from the research information which is available. This may be achieved, in a method devised by the British Welding Research Association, by consulting a table or chart in which the preheating temperature is given as a function of:-

(1) the electrode diameter,
(2) an index of weldability, and
(3) a thermal severity number.

The electrode diameter needs no further discussion but the meaning of the other factors must be clarified. The index of weldability is one of seven letters (A-G) and is based on a carbon equivalent defined by:-

$$C_{eq} = \%C + \frac{\%Mn}{20} + \frac{\%Ni}{15} + \frac{\%Cr+\%Mo+\%V}{10} \qquad (5.25)$$

This index is not unique but varies somewhat according as the electrode is basic or rutile.

The thermal severity number is a function primarily of the number of paths through which the heat may be conducted away, e.g. 2 in the case of a lap or butt weld, 3 in the case of a fillet-welded T joint, etc. The effect of plate thickness is acknowledged in a secondary way by multiplying the number of heat paths by a step-function, i.e. plate thickness introduces groups rather than a continuously variable effect. This is essential where the ultimate data are presented in tabular, as distinct from graphical, form.

SÉFÉRIAN[58] has adopted a rather different approach. He proposes a carbon equivalent:-

$$C_{eq} = \%C + \frac{\%Mn+\%Cr}{9} + \frac{\%Ni}{18} + \frac{7(\%Mo)}{90} \qquad (5.26)$$

and a thickness equivalent:-

$$C_t = 0.005e.C_{eq} \qquad (5.27)$$

where e is the plate thickness in mm.

The total equivalent C is then:-

$$C = C_{eq} + C_t \qquad (5.28)$$

and the preheating temperature θp is given simply as:-

$$\theta_p = 350\sqrt{(C - 0.25)} \qquad (5.29)$$

Applying these two methods to a 1.2% Mn-0.5% Mo steel, used in nuclear power work, gave the following preheating temperatures:-

B.W.R.A. method 175°C;   Séférian's method 190°C

The results are in reasonable agreement, bearing in mind the artifices which must be employed in making the calculations simple and quick.   Séférian's is probably slightly more suspect because it ignores the number of paths by which heat is conducted away.

Some interesting work on cooling rates was completed by the British Welding Research Association and the results disclosed at an Institute of Welding Seminar on Welding Metallurgy in 1962.   Carbon steels which contained almost nothing but iron and carbon were heated to 1000 °C and cooled to 100 °C in 7, 45, or 230 sec.   The effects of these cooling rates on hardness and microstructure of three carbon steels (0.15% C, 0.42% C, and 0.77% C) were studied and compared with the results from actual welds with the same compositions.

In the 0.15% C steel, the lowest cooling rate yielded a microstructure of free ferrite in Widmanstätten and the increases in cooling rate diminished the amount of free ferrite and led to the precipitation of carbide in extremely fine form. With the 0.42% C steel, the ferrite had disappeared by the time the highest cooling rate was achieved, the structure then being mainly martensite with odd patches of troostite.   The 0.77% C steel was troostitic at the lowest and intermediate rates and fully martensitic at the highest rate.

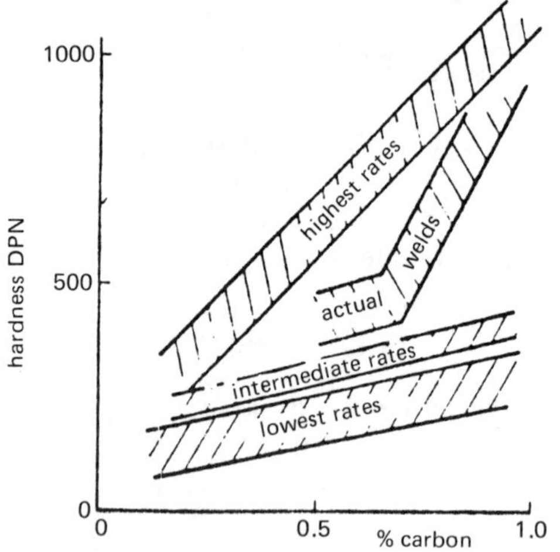

Fig. 5.24 Hardness as a function of carbon content
for actual welds and simulative tests

The hardness results showed the inevitable scatter, so that

little is lost in the diagrammatic representation of Fig. 5.24. One thing is clear − the zone of points at the highest rates has a much steeper slope than the zones for the intermediate and lowest rates; this seems to occur because the VPN never exceeds 400 if there is some troostitic structure and consequently, at these rates, the effect of cooling rate can be offset by a change in carbon content. Fig. 5.24 also indicates that the rate of cooling in an actual weld is a little greater than the intermediate rate (an average of 20 C°/sec) until the carbon content is about 0.7 per cent, after which point the hardnesses in actual welds seem to approach those of the highest rates in the cooling tests. Clearly, even with very pure alloys, there are still unexplained aspects of the possible correlation between actual welds and simulative cooling tests.

Porosity is associated with gases, although the existence of a gas does not necessarily cause porosity. The production of gases is one of the possible results of the chemical reactions which occur in the weld pool and it is timely at this point to say a little about this.

The effective temperature limit for reactions is probably about 2000 °C because it is only in the weld pool that the metal and slag vaporizes. The main elements which may be present in the weld pool are carbon, chromium, manganese, molybdenum, nickel, phosphorus, silicon, and sulphur. The extent to which any oxidation reaction takes place is governed by the free energy of formation (which is largely a linear function of temperature), i.e. if two elements are present in any oxygen (or partly-oxygen) atmosphere, the tendency is for the oxidation of that element which has the lower free energy of formation. The word 'tendency' is used advisedly, for if the element with the higher free energy of formation is present in great quantity, some of it must perforce oxidize, despite its higher energy requirements.

Information on free energy of formation is thus found to be very helpful in interpreting weld-pool reactions and, fortunately, standard diagrams do exist. Fig. 5.25 shows a free energy of formation/temperature diagram for the elements of interest and, subject to the caution mentioned previously, the reactions corresponding to the lower lines will proceed in preference to those pertaining to the higher lines, e.g. the reaction:-

$$2FeO + Si = SiO_2 + 2Fe \qquad (5.30)$$

will proceed because the SiO_2 line is appreciably lower than the FeO line.   Again, an abundance of FeO and a small amount of silicon would tend to inhibit the reaction.

All chemical reactions require time, and consequently it is at the lower temperatures (where the time available is not necessarily enough for the reaction to take place) that the diagram should be used with most caution.   Fortunately, welding temperatures are usually fairly high.

From this diagram (Fig. 5.25) the effects of various added elements can be interpreted.   For example, any FeO will be reduced to iron by carbon which will itself then become oxidized and be a potential source of porosity trouble.

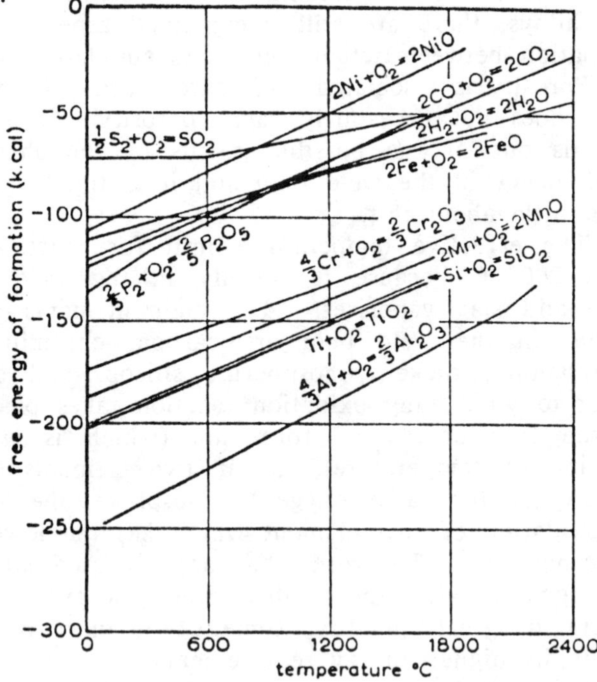

Fig. 5.25 Free-energy of formation of
oxides as a function of temperature

The addition of silicon will inhibit this by encouraging the reaction of Equation 5.30 above, in preference to the oxidation of carbon.   Similar remarks apply to manganese and the important point is that the oxides of silicon and manganese exist as comparatively harmless inclusions in the weld metal.

Again, the reaction:-

$$Ti+O_2 = TiO_2 \qquad\qquad (5.31)$$

calls for only a moderate free energy and represents a real difficulty in getting the titanium into stainless steel weld metal. Such a reaction must be inhibited and this can only be achieved by adding an element which:-

(1) is even lower on the diagram, and
(2) does not form a troublesome oxide.

Aluminium is the best for this and, although not entirely successful, it does greatly increase the amount of titanium which is present in the weld metal.

It would be inappropriate to terminate a discussion of the effects of welding without referring to the work of SCHAEFFLER[59] and, in particular, to the Schaeffler diagram. This diagram is of great importance in the welding of nickel-chrome steels in general and stainless steels in particular. There have been previous diagrams in which metallurgical structure was shown as a function of the nickel and chromium contents, but the virtue of Schaeffler's work is the concept of the *equivalent* nickel and chromium contents defined by:

equivalent nickel%   = % Ni+30 × % C+0.5 × % Mn
equivalent chromium% = % Cr+% Mo+1.5 × % Si+0.5 × % Nb.

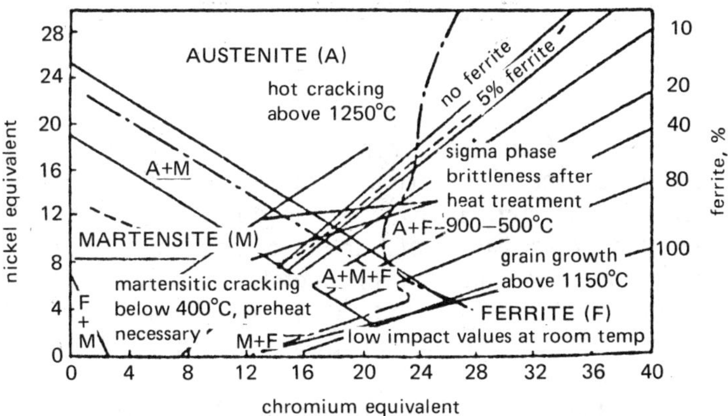

Fig. 5.26 The Schaeffler diagram (for stainless steel)[60]
(by courtesy of the Institute of Welding)

The Schaeffler diagram is a graph with equivalent nickel as ordinate and equivalent chromium as abscissa, structures and defects being indicated in terms of straight lines and areas (the diagram for stainless steel is shown in Fig. 5.26). The use of straight lines is justified mainly on the score of simplicity, but their approximation to the truth in some circumstances should be borne in mind when considering border-line cases, e.g. the presence or absence of a small amount of ferrite. Notwithstanding this, the Schaeffler diagram is of great use in the study and practice of welding Ni-Cr steels.

It is hoped that the approach to this section has made sense to the reader; perhaps it is helpful to summarize in concluding the section. The heat generation characteristics are discussed broadly in Chapter 5 of the companion volume[16]; the present chapter has sought to show the temperature levels reached (and their distribution), the heat flow patterns and cooling rates; to relate them to known metallurgical behaviour at comparable temperature levels and cooling rates in order to predict the results of welding a given material in a known manner.

An approach of this nature is always somewhat disheartening at first; there seem to be insufficient data to predict the behaviour of any one material — indeed, the approach seems to do little but expose huge gaps in knowledge. This is, in fact, a healthy sign, but it should lead to a desire to fill in the gaps and not to return to the *ad hoc* approach which necessitates each new material being studied as though it were a new problem.

5.7 Heat Treatment

The effects of heat treatment on the functioning of materials are so important that heat treatment processes can be singled out and considered specifically in that context. A detailed discussion of the various metallurgical treatments which can be imposed on metals and alloys has already been given in Chapter 2, but it is worth while mentioning them again here as production processes which have an effect on final properties.

Perhaps the most important heat treatment is that of annealing. Deformation processes break up and distort the crystal structure and lead to work hardening, especially if the final working is carried out at fairly low temperature. Continual working thus becomes more and more difficult as the metal becomes harder and larger areas of metal have to be deformed. The available ductility is lowered until fracture becomes a danger. When this happens an annealing treatment can be imposed on the metal, and its softness and ductility restored before further working. The opportunity is usually taken at this stage to produce the grain size which is satisfactory for the deformation processes which are to follow. A balance has to be struck between the final properties required and the softness needed for deformation to be accomplished with the available equipment. In the manufacture of large numbers of articles an even more important consideration is the overall optimization of the processing sequence on an economic basis, for annealing treatments are expensive, requiring large furnace capacities and long interruptions of the whole sequence of operations.

Annealing is also used to relieve residual stresses which have been induced in a component, either by prior deformation or by prior heat treatment. It should be borne in mind that a prolonged annealing treatment will always relieve residual stresses, so that if, for example, compressive residual stresses have been purposely introduced into the surface layers of a component to improve fatigue strength, an annealing treatment may be undesirable.

The process of intentionally partially annealing work hardened metals is referred to as 'temper-annealing' or 'temper-letting-down'. As these names imply, the partial annealing of work hardened metals gives a method of achieving a particular temper, as an alternative to temper-rolling, for example. This method is more often used for aluminium alloys than other metals, the actual details of treatment being empirical, each manufacturer having to find the optimum conditions by trial and error. In Table 5.3 are shown some typical temper-annealing treatments for these materials.

It will be noticed from this table that the process is not easy to control, since only 5 C° temperature difference is necessary in the case of the softer alloys to change the temper from $\frac{1}{2}$-hard to $\frac{1}{4}$-hard. Also, large masses of metal in the form of big coils or stacks of sheets cannot be brought to the desired temperature quickly, and some annealing will take place during the gradual rise of temperature. There will also be some variation in final properties throughout the furnace, due to temperature variations.

The combination of properties in temper-annealed aluminium sheet is somewhat different from that of temper-rolled sheet. For the same U.T.S. in both cases, the proof stress will be lower and elongation higher in the temper-annealed metal. Economically, there is often a small advantage in temper-annealing since the extra rolling pass required for temper-rolling is thereby avoided. On the other hand, temper-rolled sheet often has a better *shape* and *surface finish*.

The *hardening* of metallic alloys is a heat-treatment process which is almost as important as annealing. The mechanism whereby hardening takes place may be by the precipitation of a harder phase, following a quenching treatment which has resulted in either a supersaturated solid solution or the retention of a crystal structure of the metal which is only stable for long periods at higher temperatures.

TABLE 5.3

Material	Desired temper	Starting temper	Temper-letting-down conditions		
			Desired temperature: °C	Time at temperature: h	Approx. total time cycle: h
99% pure Al	$\frac{1}{2}$-hard $\frac{1}{4}$-hard	Hard-rolled (no anneals)	230 235	3	12
Al-1$\frac{1}{4}$% Mn alloy	$\frac{1}{2}$-hard $\frac{1}{4}$-hard	Homogenized Hard-rolled (no anneals)	290 295	3	12
Al-2$\frac{1}{4}$% Mg alloy	$\frac{1}{2}$-hard $\frac{1}{4}$-hard	Hard-rolled (after anneal) 50-55% reduction	175 245	1	7-8
Al-3$\frac{1}{2}$% Mg alloy	$\frac{1}{2}$-hard $\frac{1}{4}$-hard	Hard-rolled (after anneal) 50-55% reduction	110 235	1	7-8

From FIELD *and* SALTER[61]

The supersaturated solid solution is often quite soft and easily worked, the final hardness being achieved by the gradual precipitation of a harder phase, which may be accelerated by heating. On the other hand, a metal which on quenching retains an allotropic form normally stable only at high temperature, may be extremely hard and brittle. Subsequent heating may then be necessary, not to accelerate age-hardening, but to temper the quenched structure.

It is impossible in a discussion of heat treatment to

disregard possible deleterious effects of the process on the metal. An example has already been mentioned in connection with temper-annealing in that in some types of industrially supplied furnace there will be variations of temperature which will inevitably result in variations of properties throughout the volume of the product, unless care is taken to stack the individual pieces suitably. Also, unless the chemical composition of the furnace atmosphere is carefully controlled and/or the surface of the metal protected, chemical reactions will take place such as oxidation, decarburization, and dezincification. Another possible defect which may arise from heat-treatment is blistering, due to the expansion of trapped gases which were either present in the metal itself or caused by reaction with furnace gases.

An extensive discussion of the *control of quality* in the production of wrought *non-ferrous* metals and alloys was held by the Institute of Metals over 30 years ago, and this gave a valuable account of practice in many industries[61], which is still relevant. Previously, the same Institute had held a symposium in which equipment for heat treatment was considered[62]. Both of these references give considerable detailed information on the subject, especially concerning the alloys of aluminium, copper, and nickel. Another symposium held by the American Society for Metals[63] and the book by JENKINS[64] both discuss controlled atmospheres.

To produce a properly controlled atmosphere for simple carbon steels, the work must be enveloped by a suitable gas in an enclosed chamber. An inert gas such as helium or one inert to most steels, such as nitrogen, could theoretically be used but in practice it is very difficult to avoid the ingress of air. Water vapour can be formed by the reaction between CO_2 and H_2, and it is extremely oxidizing and decarburizing to steel. Thus a suitable gas would be one containing only N_2, CO, and CO_2 in which the CO/CO_2 ratio has been controlled to avoid oxidation at the particular temperature concerned. Similarly a gas containing only N_2, H_2, and H_2O can be used providing the H_2/H_2O ratio is controlled. Alloy steels behave similarly to carbon steels but are more prone to decarburization and will therefore stand less deviation from the equilibrium composition of the furnace atmosphere. Since Cr is oxidized by CO, CO must be kept well away from high chromium steels. For stainless steels an atmosphere of H_2 or N_2 and H_2 with no trace of water is essential, to avoid staining.

This brief account of some of the problems which arise

in the heat treatment of metals, in the context of the effect on the functioning of the product, reveals that the subject is too vast for any attempt to be made to cover it comprehensively in a book of this size. The importance of the correct control of furnace type, temperature, atmosphere, heating and cooling rates, and the method and geometry of stacking have been emphasized. Established procedures for particular metals and alloys have been painstakingly developed over the years and should be ascertained and adhered to. For example, it is quite wise to reserve one particular furnace for one particular alloy if possible.

5.8 Discussion

In this chapter an attempt has been made to identify some of the considerations influencing the choice of operations required to produce a certain component in a given material, relating particularly to the effect of the process on the functioning of the material in the product. Wider issues are involved in the actual choice of material however; for example, design considerations such as specification of the desired life of the artefact, its resistance to corrosion and wear, necessity for maintenance, safety of operating personnel, and the over-all economics of the manufacture of the complete product, including marketing factors.

For example, consider the problem of manufacturing a specific item, an automobile engine connecting rod. This is a highly stressed fast-moving and important component which would cause considerable damage if it failed in service. A high-strength aluminium alloy is an obvious choice for the designer to make and he knows the imposed loads and approximate shape of the part. The fatigue strength of this component is obviously of paramount importance.

The main methods of manufacture which can be considered for this part are casting, forging, powder metallurgy, or machining from a solid forged billet. Large numbers of the component are required, all to carefully controlled dimensions. Machining from solid bar stock on modern CNC machining centres is attractive, particularly as modern CNC and CAM techniques enable the easy and accurate generation of the complex shape required. That process would result in considerable waste of the material, however, and the metallurgical structure of the final product would not be even near to the optimum required for its highly stressed application.

Casting would also be attractive from the point of view of obtaining the final shape easily and quickly with the minimum waste of material. As pointed out in Chapter 1, the forming of liquid or near-liquid metal is theoretically the most efficient method available. The final cast metallurgical structure would generally have inadequate strength for this particular application and it would be difficult to make a con-rod of any reasonable size which would have the required fatigue strength. However, in recent years, some automobile con-rods _have_ been cast in spheroidal cast iron on a successful production scale, having sufficient fatigue strength for their particular duty.

Powder metallurgy has similar attractions, as does casting from the point of view of economizing in material. One attractive such material is SAP (Sintered Aluminium Powder), which possesses high resistance to creep and to mechanical wear but rather low fatigue strength, unfortunately.

For this particular application forging has always been the preferred method, the most economical sequence being to hammer forge hot from bar stock in the early stages so that the metal can be quickly brought to the rough final shape, finishing by cold forging to produce a good surface finish with accurate dimensions. In this way a suitably orientated metallurgically fine-grained structure can be introduced into the metal, giving maximum fatigue strength.

Another important example from modern technology is that of the manufacture of gas turbine components such as rotors, discs and spools (thin cylindrical tubular rotors). Although these are all essentially axi-symmetrically shaped parts, they are often required to be of thin awkward shape making the part difficult, if not impossible, to forge and to remove from between the dies, which often have to incorporate re-entrant shapes preventing removal of the part. The effect of the production process on the functioning of the final artefact is such an important consideration in the case of these components in which extremely high stresses are induced in hostile environments, that they have at present to be manufactured by massive bulk machining from solid forged billets of high initial strength and integrity. Because of this, as much as 97% of the billet, generally made from expensive metallic alloys of nickel, titanium, or stainless steel, for example, is machined away to become swarf. The resultant swarf is recycled whenever possible but the actual cost of manufacturing is enormously increased because of the over-all consumption of expensive energy involved in recycling such expensive metals and it is often not feasible anyway.

The whole problem of the manufacture of these 'space-age' components by *nett shape forming* or *near to nett shape forming* methods forms the subject of current research in many industries and universities, internationally. The development of new processes for joining parts together and building up components of this type by processes such as *deposition forming* has been discussed by ALEXANDER[65]. One of the main problems is to achieve the same strength against fatigue, creep and fracture as can be achieved by the present methods of, essentially, bulk machining the parts from solid forged billets. The process of *friction welding*, sometimes referred to as *inertia welding*, discussed elsewhere[16], is also a very attractive method for the high-strength bonding together of individual components, particularly of axi-symmetric shapes like the gas turbine engine spool, to form a final complex shape.

Each case must be considered on its own merits and it is difficult to lay down any firm set of guiding principles. To a large extent the principles involved must form part of a broader philosophy of design resting on a sound basis of economic appraisal of the whole problem of producing an engineering structure capable of performing a certain function for a certain lifetime within given limits of accuracy and safety. Viewed in this wider context, the choice of the material to be used assumes paramount importance – it is pointless choosing stainless steel to manufacture an article for which a cheap plastic could be made adequate by suitable design modifications.

REFERENCES

(1) POLUSHKIN, E.P., *Defects and Failures of Metals.* Elsevier, Amsterdam, 1956.

(2) FLEMINGS, M.C., *Solidification Processing.* McGraw-Hill, 1974.

(3) WIESER, P.F. (Ed.) *Steel Castings Handbook.* Steel Founders Soc. Amer., (5th Ed.), 1980

(4) CAMPBELL, J. and WILKINS, P.S.A., "New Developments in Light Alloy Founding", *Brit. Foundrymen* 1982,75,233-236.

(5) RUDDLE, R.W., *The Solidification of Castings.* Inst. of Metals Monograph No.7, May 1957.

(6) DUMA, J.A. and BRINSON, S.W., "Application of Controlled Directional Solidification to Large Steel Castings", *Trans. Amer. Found. Assoc.* 1940,48,225.

(7) JACKSON, R.S., "Application of Insulated Feeders to Sand Castings in Long Freezing Range Copper Alloys", *Found. Trade J.* 1956,100,487.

(8) CIBULA, A., "Influence of Grain Size on the Structure, Pressure Tightness and Tensile Properties of Sand-Cast Bronzes and Gun-Metals", *Proc. Inst. Brit. Foundrymen* 1955,48,A73.

(9) RUDDLE, R.W., "Correlation of Tensile Properties of Aluminium Alloy Plate Castings with Temperature Gradients during Solidification", *J. Inst. Metals* 1950,77,37.

(10) WALTHER, W.D., ADAMS, C.M. and TAYLOR, H.F., "Better Aluminium Castings", *Modern Metals* 1954,10,44.

(11) BRIGGS, C.W., GEZELIUS, R.A., and DONALDSON, A.R., "Steel Casting Design for the Engineer and Foundryman", *Trans. Amer. Found. Assoc.* 1938,46,605.

(12) BRIGGS, C.W., *The Metallurgy of Steel Castings.* McGraw-Hill, New York, 1946.

(13) BRINSON, S.W. and DUMA, J.A., "Studies on Center-Line Shrinkage in Steel Castings", *Trans. Amer. Found. Assoc.* 1942,50,657.

(14) CAINE, J.B., "A Theoretical Approach to the Problem of Dimensioning Risers", *Trans. Amer. Found. Assoc.* 1948,56,492; 1949,57,66.

(15) FORTINO, D., "Sulla Determinazione della Tendenza alle Tensioni Residue della Ghisa Neigetti", *La Fonderia Italiana* 1960, No.3, 89.

(16) ALEXANDER, J.M., BREWER, R.C. and ROWE, G.W.,
Manufacturing Technology, Vol.2: Engineering Processes.
Ellis Horwood, Chichester, 1986.

(17) OLIVER, D.A., "Heat Resistant Materials: a Survey of
some British Developments",
Proc Int. Congr. Aero. Soc. Madrid, 1958.

(18) GREGORY, E. and GOETZEL, C.G., "Some
High-Temperature Properties of Nickel Alloy Powder
Extrusions Containing Non-Metallic Dispersions",
Trans. Amer. Inst. Min. and Metall. Engrs. 1958,212,868.

(19) BRUCKART, W.L., CRAIGHEAD, C.M. and
JAFFEE, R.I., "Investigations of Molybdenum and
Molybdenum Base Alloys Made by Powder Metallurgy
Techniques", *W.A.D.C. Tech. Rep.* 54-398, 1955.

(20) GLENNY, E and TAYLOR, T.A., "High Temperature
Properties of Ceramics and Cermets",
Powder Metallurgy 1958, No.2, 189.

(21) BREWER, R.C. and PEARSON, J.H., "Some
Considerations of Toughness in Sintered Tool Materials",
Int. J. Mach. Tool Design and Res. 1962,3,25.

(22) PARIKH, M.N. and HUMENIK, M., "Cermets",
J. Amer. Ceram. Soc. 1957,40,315,335.

(23) ROYSTON, M.G. and BARRETT, L.R., "Some
Observations and Reservations on Thermal Shock Theory",
Trans. Brit. Ceram. Soc. 1958,57,678.

(24) WHITE, J., "Some General Considerations on Thermal
Shock", *Ibid.* 591.

(25) CADDELL, R.M., *Deformation and Fracture of Solids.*
Prentice-Hall, 1980.

(26) JONES, W.D., *Fundamental Principles of Powder
Metallurgy.* Edward Arnold, London, 1961.

(27) RICHARDS, C.E. and LYNCH, A.C., (Eds.) *Soft
Magnetic Materials for Telecommunications.*
Pergamon, London, 1953.

(28) SMIT, J. and WIJN, H.P.J., *Ferrites.*
Philips Technical Laboratory, Eindhoven, 1959.

(29) SCHALLERER, J.W., "Latest Developments in Magnetic
Storage and Switching Applications",
Proc. 13th Ann. Mtg. Metal Powder Assoc. 1957,11

(30) SCHEIDEGGER, A.E., *The Physics of Flow Through
Porous Media.*
University of Toronto Press (2nd Ed.), 1962.

(31) MORGAN, V.T. and CAMERON, A., "Mechanism of
Lubrication in Porous Metal Bearings", *Proc. Conf. Lub.
and Wear,* Paper No. 89, 1957, I.Mech.E.

(32) WOLDMAN, N.E., *Metal Process Engineering*.
Reinhold, New York, 1948.

(33) SACHS, G., *Fundamentals of the Working of Metals*.
Pergamon, London, 1954.

(34) DODD, B., "Shear Instabilities in Blanking and Related
Processes", *Metals Technology*, 1983,10,57-60.

(35) PEARSON, C.E., *The Extrusion of Metals*.
Chapman and Hall, London (2nd Ed.), 1960.

(36) CLAPP, B.P., "Economics of Cold Extrusion",
National Engineering Laboratory Report 1965,**195**.

(37) CARSLAW, H.S. and JAEGER, J.C., *Conduction of Heat
in Solids*. Clarendon Press, Oxford, 1959.

(38) WEINER, J.H., "Shear Plane Temperature Distributions in
Orthogonal Cutting", *Trans.A.S.M.E.* 1955,77,1331.

(39) RAPIER, A.C., "A Theoretical Investigation of the
Temperature Distribution in the Metal Cutting Process",
Brit.J.App.Phys. 1954,5,900.

(40) BOOTHROYD, G., *Fundamentals of Metal Machining*.
Edward Arnold, London, 1965.

(41) SHIRAKASHI, T. and USUI, E., "Simulation Analysis of
Orthogonal Metal Cutting Mechanism",
Proc.Int.Conf.Prod.Eng., Tokyo 1974,p535.

(42) STEVENSON, M.G., WRIGHT, P.K. and CHOW, J.G.,
"Further Developments in Applying the Finite Element
Method to the Calculation of Temperature Distributions
in Machining", *A.S.M.E.,J.Eng.Ind.* 1983,105,149-154.

(43) TRENT, E.M., *Metal Cutting*. Butterworths, 1977.

(44) SMART, E.F. and TRENT, E.M., "Coolants and Cutting
Tool Temperatures",
Proc.15th Int.M.T.D.R.Conf., 1975,pp187-195.

(45) CHAO, B.T. and TRIGGER, K.J., "Temperature
Distribution at Tool-Chip and Tool-Work Interfaces in
Metal Cutting", *Trans.A.S.M.E.* 1958,80,311.

(46) EL-HELIEBY, S.O.A. and ROWE, G.W., "Influences of
Surface Roughness and Residual Stress on the Fatigue
Life of Ground Steel Components",
Metals Technology 1980,7,221-225.

(47) MITTAL, R.N. and ROWE, G.W., "Residual Stresses and
their Removal from Ground En31 Steel Components",
Metals Technology 1981,9,191-197.

(48) SPRINGBORN, R.K. (Ed.), *Non-Traditional Machining
Processes*. Soc.Mech.Engrs., 1967.

(49) *Source Book on the Application of the Laser in
Metalworking*. A.S.M., 1979.

(50) *Tool and Manufacturing Engineers' Handbook.*
McGraw-Hill, New York, 1976.

(51) CHILDS, T.H.C. and ROWE, G.W., "Physics in Metal
Cutting", *Rep.Prog.Phys.* 1973,**36**,223-288.

(52) ROSENTHAL, D., "Mathematical Theory of Heat
Distribution during Welding and Cutting",
Welding J. 1941,**20**,2205.

(53) HANEMANN, H. and SCHRODER, A., *Atlas
Metallographicus.* Vol. III, p. 30.
Verlag Stahleisen, Dusseldorf, 1952.

(54) TIPPER, C.F., *The Brittle Fracture Story.*
Cambridge University Press, 1962.

(55) VAUGHAN, H.G. and MORTON, M.E., "Hydrogen
Embrittlement of Steel and its Relation to Weld Metal
Cracking", *Brit. Weld. J.* 1957,**4**,40.

(56) FLANIGAN, A.E. and MICLEU, T., "Relation of
Pre-Heating to Embrittlement and Microcracking in Mild
Steel Welds", *Weld. J.* 1953,**32**,99.

(57) COTTRELL, C.L.M., "Controlled Thermal Severity
Cracking Test Simulates Practical Welded Joints",
Weld. J. 1953,**32**,257.

(58) SÉFÉRIAN, D., *Metallurgie de la Soudure.*
Dunod, Paris, 1959.

(59) SCHAEFFLER, A.L., "Constitution Diagram for Stainless
Steel Weld Metal", *Metal Progress* 1949,**56**(5),680.

(60) BORLAND, J.C. and YOUNGER, R.N., "Some Aspects
of Cracking in Welded Cr–Ni Austenitic Steels (A Survey
of the Literature)", *Brit. Weld. J.* 1960,**7**(1),22-59.

(61) FIELD, R. and SALTER, G., "The Control of Quality in
Heat-Treatment and Final Operations",
Institute of Metals Monograph and Report Series 1955,**17**,

(62) "Equipment for the Thermal Treatment of Non-Ferrous
Metals and Alloys", *Institute of Metals Monograph
and Report Series* 1953,**14**,

(63) "Controlled Atmospheres",
American Society for Metals Publication, 1941.

(64) JENKINS, I., *Controlled Atmospheres for the Heat-
Treatment of Metals.* Chapman and Hall, London, 1946.

(65) ALEXANDER, J.M., "The Economics of Alternative
Methods of Metal Processing", *Annals C.I.R.P.* 1984,**33**(2),

AUTHOR INDEX

The names of authors are followed by the numbers of the pages where there is a reference to them or to their published work. Numbers within parentheses refer to the same publication, the italic number indicating the page on which the complete reference may be found.

SUBJECT INDEX